Lecture Notes in Chemistry

Edited by G. Berthier, M. J. S. Dewar, H. Fischer
K. Fukui, H. Hartmann, H. H. Jaffé, J. Jortner
W. Kutzelnigg, K. Ruedenberg, E. Scrocco, W. Zeil

19

Enrico Clementi

Computational Aspects for
Large Chemical Systems

Springer-Verlag
Berlin Heidelberg New York 1980

Author

Enrico Clementi
IBM Data Processing, Product Group, P.O.Box 390
Poughkeepsie, New York, 12602/USA

ISBN-13: 978-3-540-10014-0   e-ISBN-13: 978-3-642-93144-4
DOI: 10.1007/978-3-642-93144-4

Library of Congress Cataloging in Publication Data. Clementi, Enrico. Computational aspects for large chemical systems. (Lecture notes in chemistry; 19) Includes bibliographical references and index. 1. Quantum chemistry--Mathematics. I. Title. QD462.C53 541.2'8'0151 80-19390

This work is subject to copyright. All rights are reserved, whether the whole or part of the material is concerned, specifically those of translation, reprinting, re-use of illustrations, broadcasting, reproduction by photocopying machine or similar means, and storage in data banks. Under § 54 of the German Copyright Law where copies are made for other than private use, a fee is payable to the publisher, the amount of the fee to be determined by agreement with the publisher.

© by Springer-Verlag Berlin Heidelberg 1980
Softcover reprint of the hardcover 1st edition 1980
2152/3140-543210

FORWARD

Basic limitations are inherent in those attempts aimed at a description of complex chemical systems (like bio-chemical systems) if only quantum chemical methods (namely, with methods developed mainly to obtain approximated solutions to the Schroedinger equation) are adopted. One of such limitations is apparent when as "elementary objects" nuclei and electrons rather than groups of atoms are considered. A second set of limitations is the neglect of temperature and time parameters. Essentially, in much of today's quantum-chemistry insufficient attention has been paid to establishing an operational link to statistical dynamics and thermodynamics. Concepts and related computational techniques are presented to establish such a link more firmly. In the first four chapters, we deal mainly with quantum mechanics related to large systems. From the start we stress complex rather than simple molecular system. In the fifth chapter we consider the statistical Monte Carlo method giving examples for ions, ions pairs, amino acids, DNA's and RNA's separated bases, base pairs, up to DNA (single and double helix) in solution. The solvent effect in enzymes is analyzed in two examples, lysozyme and carbonic anhydrase. We feel that such examples, even if representing some of the most advanced computation today available will become routinely within few years.

One of the aims of this work is to show that computational chemistry is neither limited to simple systems nor to unphysically gross approximations when it deals with complex chemical systems. A second goal is to provide an operational framework of concepts and algorithms to describe matter either at the electronic level or at the atomic and molecular level and (later) at the continuum level; in this way starting with quantum chemistry, we move to statistical mechanics then to thermodynamics and (later) to fluid mechanics.

I wish to thank the collaborators of a number of unpublished studies reported in this work; in particular I wish to thank Drs. G. Corongiu and S. Romano. In addition it is my pleasure to thank Drs. G. Corongiu, L. Gianolio, R. Pavani, R. Ranghino, C. Roetti and S. Romano and Profs. W. Kolos and O. Novaro for the collaboration at papers, presently in press, and partially reported in these notes.

A very preliminary version of this report was distributed at the 14th meeting of the Italian School on the Structure of Matters sponsored by the National Research Council of Italy (University of Perugia, Department of Physics, 27th of August - 8th of September, 1979).

These notes have been used as the main material for the "Firth Visiting Professor", 1980 lecture series at the University of Sheffield, (Sheffield, U.K.) and for the E. H. Boomer Memorial Lectures, 1980, at the University of Alberta (Edmonton, Canada).

I dedicate these Notes to my postdoctoral period's advisors namely Prof. Giulio Natta, Politecnic Institute of Milan, Department of Industrial Chemistry (Milan-Italy 1955), Prof. Michael Kasha, Florida State University, Department of Chemistry (Tallahassee, Florida, U.S.A., 1956-1957), Prof. Kenneth S. Pitzer, University of California, Department of Chemistry (Berkeley, California, U.S.A., 1958 and part of 1959), Prof. Robert S. Mulliken, University of Chicago, Department of Physics (Chicago, Illinois, U.S.A., part of 1959 and 1960) and to Dr. Arthur G. Anderson (International Business Machines Corporation) for his continuous support, encouragement and advice during the last twenty years.

Finally it is my pleasure to thank the Internal Documentation Processing Department (IBM, Poughkeepsie) and Dr. G. Corongiu for their help and collaboration in preparing the manuscript.

CONTENTS

1.0	INTRODUCTION	1
1.1	Statement of the Problem	1
1.2	Definition of Chemical Complexity	2
1.3	On the Upper Limit of Quantum Chemical Computations	3
1.4	A General Method for Simulations of a Complex Chemical System	13
2.0	COMPLEXITY BECAUSE OF THE "SIZE" OF THE LARGEST MOLECULE IN THE SYSTEM	17
2.1	Comments on Conformational Analyses for a Single Molecule	17
2.2	A New Method for Protein-Substrate Interaction Simulations	18
	2.2.1 Macrodeformations	21
	2.2.2 Microdeformations	22
2.3	Further Improvements for Enzymatic Reaction Simulations	26
3.0	ANALYSES OF CHEMICAL BONDS	28
3.1	Introduction	28
3.2	Bond Energy Analysis	29
3.3	One-Center Energies and the Molecular Orbital Valance State	31
3.4	Two-Center Bond Energy: Benzene	32
3.5	Orbital and Electron Energies	33
3.6	MOVS and Hybridization	36
3.7	Bond Energy Analysis: A "New" Formalism	37
3.8	Chemical Formulae From the Bonded Atom Pairs Analysis	41
3.9	Definition of Atoms and Molecules	44
3.10	BEA and Reaction Surface	48
3.11	BAP and Reaction Surface	52
3.12	Bond Energy Analysis and Vibrational Analysis	58
4.0	ATOM-ATOM PAIR POTENTIALS	59
4.1	Preliminary Comments	59
4.2	Atomic Classes for Atoms in Molecules	59
4.3	Determination of Two-Body Pair Potentials	62
4.4	Pair Potentials and Ab Initio Computations	63
4.5	Minimal Basis Set and Basis Set Superposition Error	64
4.6	The Dispersion Energy	67
4.7	Three and Many Body Corrections	78
5.0	COMPLEXITY BECAUSE OF THE NUMBER OF COMPONENTS IN THE CHEMICAL SYSTEM	86
5.1	Liquid Water	86
5.2	Ion Water Clusters: Two Body Potentials	89
5.3	Ionic Solutions: Effective Two Body Potentials	90
5.4	Ionic Solutions: n-Body Correction	92
5.5	Energy Maps and Water Structure In Solutions	92
5.6	Monte Carlo Simulation of the Interaction Between Glycine and the Corresponding Zwitterion	99
5.7	Serine and the Corresponding Zwitterion	104
5.8	Enzyme-Water Interaction in Solution: A Preliminary Study on Lysozyme	114
5.9	The Water Structure in the Active Cleft of Human Carbonic Anhydrase/B	124
5.10	Contour Maps for the Molecular Fragments of DNA	131
5.11	Monte Carlo Simulations for Bases and Base-Pairs in Nucleic Acids	143
5.12	Solvation of B-DNA Double Helix at $T=\underline{300}°K$	152
5.13	Solvation of Na^+-B-DNA at $\underline{300}°K$	167
5.14	Conclusion	176
6.0	REFERENCES	178

1.0 INTRODUCTION

1.1 STATEMENT OF THE PROBLEM

Quantum chemistry judged not from the ever present possibility of unexpected developments but on the basis of the achievements in the last fifty years, is predominantly limited to attempts to solve for the energy and expectation values of wave functions representing, in the limit, an exact solution to the Schroedinger equation. Because of well-known difficulties in system with more than about 50 electrons, the adopted approximations are generally rather crude.

As examples of quantum chemical approximations we mention the total or partial neglects of electron correlation, the neglect of relativistic effects, the use of subminimal basis sets, the still present neglect of inner-core electrons in semi-empirical methods, the acceptance of the Born-Oppenheimer approximations, and so on. In general, the larger the system, in terms of the number of electrons, the cruder the approximation. In a way, the present status of quantum chemistry might appear as nearly paradoxical. Indeed, for small systems, where very accurate experiments are often available, and therefore, there is not a great need to obtain (from quantum chemistry) predictions of new data but rather a theoretical interpretation of the existing data, we find increasingly powerful and reliable quantum chemical methods and techniques. However, for complex chemical systems like, for example, those typical in catalyses and in biochemistry where the experimental data is often available in such a form as to need unambiguous interpretation from theoretical models and where there is need for quantitative predictions, we often find a diffuse use of rather unreliable quantum chemical methods and techniques. On the positive side, one can notice a greater and more diffuse awareness to the above problem: whereas only ten years ago quantum biology was limited to the CNDO type of approximations and twenty years ago the Hückel approximation was either presented or used as reliable (despite strong evidence to the contrary obtained from studies on small systems), today an increasing number of researchers realize that the intrinsic importance of a biological problem is not a sufficient justification for the use of unreliable quantum chemical approximations. In addition, a large fraction of the past work in quantum biology did consist of correlations between families of molecules, where the basic parameter was the difference in the topology of the molecules. As a consequence the use of quantum-chemical techniques was nearly irrelevant and mainly a way to obtain some correlation. However, we still note a tendency to use, for large chemical systems, those techniques developed for <u>small systems</u>, rather than attempt to search for new methods and new techniques devised explicitly to deal with <u>large systems</u>. Any model is useful within a given range of applications: outside the range it might be correct, but not necessarily useful. This obvious observation does not seem to have been appreciated by many quantum-chemists dealing with complex systems.

Likely the lack of understanding of the proper applicability range of quantum chemistry is because of the traditional neglect of parameters such as time, temperature and entropy, namely, of those parameters that since the last century have been considered as basic to describe complex chemical systems. In this context, the need to operationally link quantum chemistry to statistical thermodynamics should be obvious.

In this work, we shall consider few limited aspects concerning relations between quantum chemistry and statistical methods. Formal relations are known since the beginning of wave mechanics and therefore are not included in this work that will stress mainly operational aspects.

1.2 DEFINITION OF CHEMICAL COMPLEXITY

The complexity of a chemical system, considered as a discrete and numerable ensemble of particles, is proportional to the number of mutually dependent variables characterizing a given aspect of the system. In our implicit definition of complexity, we use two criteria:

1. <u>The number</u> of atoms or molecules present in the system and

2. <u>The size</u> of the largest molecule in the system. The size is defined as the number of atoms in the largest molecule of the system.

When one deals with biological systems, the above conditions are necessary to define the system, but probably not sufficiently, since in defining a biological system one must include conditions, whereby, for a limited interval of time, structural forms evolve from previous forms at little energy cost.

Let us consider two examples of a complex system: in the first, complexity results from the <u>number</u> of molecules, in the second, from the <u>size</u> of the largest molecule.

A single water molecule is here defined as a simple system, a cluster of about ten water molecules represents a case of intermediate complexity and liquid water is an example of a complex system. Here complexity is due to the <u>number</u> of the components. The mutually dependent variables are the positions and orientations of the water molecules. The sugar molecule (or a phosphate group), the sugar-phosphate-sugar complex, and a protein are examples of simple, intermediate, and complex chemical systems, where complexity is due to the <u>size</u> of the largest component of the system. The mutually dependent variables are the angles of internal rotations.

The two criteria, <u>number</u> and <u>size</u>, cover a large number of chemically complex systems (see Figure 1). In a simple system, the appropriate model (quantum mechanics) assumes as particles the nuclei and the electrons; thus the dominant statistics is the one of Fermi-Dirac. In complex systems the appropriate particles are atoms or group of atoms; the model is classical in the limit and the Boltzmann statistics (and time fluctuations) are essential for a proper description. It follows, as a corollary, that <u>intramolecular</u> forces play a dominant role in simple systems, whereas <u>intermolecular</u> forces, and torsional and rotational barriers are dominant in complex systems. As a second corollary, a valid goal in small systems is the determination of the exact geometrical relationship between the component atoms. On the contrary, the probabilistic nature of the distribution of the atoms or groups of atoms is of basic importance in a complex system. The notion of probability distribution brings in most naturally the need for the concept of entropy. This concept has been somewhat overlooked in quantum chemistry, since subtly built into the Schroedinger wave mechanical representation, whereas it is more easily grasped in classical chemical particles.

In conclusion, depending on the complexity of the system, one uses either quantum mechanics, statistical mechanics, or fluid mechanics and thermodynamics. In biological systems an amino acid, a protein, and a living cell provide an example of systems appropriate to the above three models.

One can safely predict that important advances in the description of complex chemical systems will be obtained by attempting to connect quantum mechanics with statistical mechanics, and statistical mechanics with fluid dynamics and thermodynamics. As noted, the need is not as much in a "formal" connection, that essentially is available, but mainly in an "operational" connection, for example, by means of computational methods.

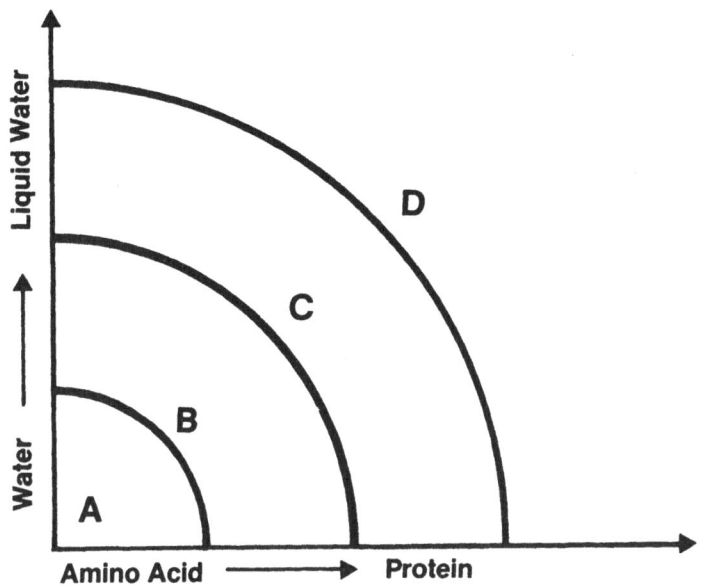

Figure 1. Simple, to intermediate, to complex chemical systems (A,B,C, respectively): the two criteria are Size of the Largest Molecule and Number of Molecules. At the limit (region D) starts the continuous description of matter.

1.3 ON THE UPPER LIMIT OF QUANTUM CHEMICAL COMPUTATIONS

We pose the question on a reasonable upper limit for quantum mechanical computations, namely we ask what is the largest number of atoms for which it is still reasonable and at times necessary to request solutions of the type available from the Schroedinger equation. We estimate that the upper limit is between 150 and 200 atoms (excluding cases of high symmetry where one might have to consider a larger number of atoms). To obtain this estimated value by computations, we have kept in mind relationships between internuclear distances and electronic density, the decay of overlap integrals, the long range nature of coulomb forces, the conformational flexibility of molecules, and other general factors of this type. Excluding the situation of systems with symmetry, to our knowledge, no ab initio computation with more than 60 to 80 atoms has been published, and for the largest computed systems, the approximations are generally at the minimal basis set level. It is therefore of practical importance to devise methods that can be reliably used for systems approaching the above limit. In our opinion there are two possible avenues; one introduces reasonable approximations at the matrix elements and at the SCF level, the second is based on new techniques for integrals evaluations and in addition it introduces the approximations at the SCF level.

One way to implement the first avenue is to partition the field, experienced by each atom of the system, into two parts: one part is obtained by the standard computation of the matrix elements and is used for small and intermediate distances, the second part makes use of electrostatic approximations and is used for large internuclear separations.

In the following we shall outline a formalism capable of handling systems of up to 200 atoms. The outline is divided into three parts starting with an analysis of the SCF problem. We recall that about twenty iterations are generally needed to reach convergence and that the storage need is related to the fact that entire matrices need to be in the fast memory of the computer. Then we consider a new way to obtain the SCF matrix elements, and finally we shall consider the electron correlation correction.

As known, the canonical set of molecular orbitals (MO) is built as a linear combination of a basis set of functions, generally with the origin at the nuclear position of the atoms. In practice, in large systems, many coefficients of the linear combination are zero, either because of symmetry or, more importantly, because the matrix elements are exceedingly small in value. We consider a macromolecule as build-up of a set of fragments, that we shall call "virtual radicals." The virtual radicals are described at first, by wave functions that are localized on the virtual radical; only in a second stage the virtual radical wave functions are combined in order to yield the wave function for the macromolecule.

For example. we could consider $CH_3-C_6H_5$ as the combination of two virtual radicals CH_3 and C_6H_5. The virtual radical's orbitals of CH_3 are constrained to be linear combinations of the basis set having the atomic functions centered on the CH_3 nuclei and similarly we built up the wave function for the virtual radical C_6H_5. Let us describe this algorithm in more detail. In the first phase, the orbitals and the wave functions for the virtual radical are computed with the above described constraints. To mitigate the physical assumptions, the virtual radical wave function is computed in the field of all the point charges of the macromolecule. As known, this brings about a very insignificant increase in computational time in properly constructed computer programs. The point charges are obtained, for example, from Mulliken's electron population analysis (1) carried out on the wave function describing both the virtual radical of interest and the two (or more) radicals contiguous to it. The Add and Merge algorithms, long ago proposed (2) and implemented (3) are most efficient in this phase of the computation. The concept of virtual radical is very much similar to the concept of virtual state; however, the external field due to point charges makes the former more physically appealing. The computational output of the first phase consists of a set of virtual radical's orbitals, all but one doubly occupied (for closed-shell macromolecules). In the second phase of the computation, the virtual radical's orbitals of the closed-shells are ortho-normalized over the entire macromolecule. Then the SCF procedure is applied only to the open-shell virtual radical's orbitals, keeping the closed-shell radical orbitals variationally frozen, and, optionally, using the open-shell orbitals as an initial guess. The procedure can be repeated as many times as reasonable, freezing once the formerly doubly occupied orbitals and once the formerly singly occupied orbitals. By construction, the matrices needed to solve the pseudo-eigenvalue equations are either small (for the virtual radical orbital) or nearly diagonalized, thus enhancing the sparse matrix characteristic of the interactions matrix elements in the macromolecule. The selection of where to cut a bond, thus creating two virtual radicals is critical: an improper choice that generally coincides with an unchemical choice, can lead to energy oscillations in the above SCF procedure. A bond containing two electrons is a reasonable one to be cut (see Section 3.8).

Let us consider, as an hypothetical example, a macromolecule composed of m atoms (m being about 200 atoms), subdivided into N virtual radicals (K=1, ..., N), each radical containing $2n'(k)+n''(k)$ electrons, $2n'(k)$ in doubly occupied orbitals and $n''(k)$ in singly occupied orbitals. The k-th virtual radical is described by the doubly occupied orbitals $\phi d'(1,k),...\phi d'(n'(k),k)$ and by singly occupied orbitals $\phi s'(1,k),...\phi s'(n''(k),k)$. The notation $\phi d'$ and $\phi s'$ as well as the following groups of symbols $\varepsilon v'_k$, F'd, F's, ϕf, ϕu, is intended to represent a single symbol; the use of more than one letter is proposed in order to limit the number of subscripts and superscripts and uncommon symbols which present some problems in computer-oriented typing and retrieval. The pseudo-eigenvalue equations are therefore of the form:

$$<\phi \mathbf{d}'(1,k)|Fd'(k,1)|\phi d'(1,k)>=\varepsilon v'(1,k)$$
$$\vdots$$
$$<\phi d'(n'(k),k)|Fd'(k,(n'(k)))|\phi d'(n'(k),k)>=\varepsilon v'(n'(k),k) \quad 1.1$$

and

$$<\phi s(1,k)|Fs'(k,1)|\phi s'(1,k)>=\varepsilon s'(1,k)$$
$$\vdots$$
$$<\phi s(n''(k),k)|Fs'(k,n''(k))|\phi s'(n''(k),k)>=\varepsilon s'(n''(k),k) \quad 1.2$$

where the upper upex indicates that the orbitals extend over the k-th virtual fragments atoms in the field of the point charges of the rest of the macromolecule. We stress that the decision on where to cut a bond in the macromolecule (to initiate a virtual fragment) can be rather crucial, especially for convergence problems in the following SCF phases. After ortho-normalization of the $\phi d'$ orbitals, one obtains a set of ϕf orbitals, that are considered variationally frozen in the next SCF phase. The total wave function of the macromolecule is in addition formed by the ϕu set of orbitals, corresponding to the ϕs from all the N virtual radicals. Thus we have orbitals obtained by solving the equations:

$$\vdots$$
$$<\phi f(1)|Ff(1)|\phi f(1)>=\varepsilon f(1) \quad 1.3$$
$$\vdots$$

and free orbitals corresponding to the equations

$$\vdots$$
$$<\phi u(t)|Fu(t)|\phi u(t)>=\varepsilon u(t) \quad 1.4$$
$$\vdots$$

where ϕf and ϕu stands for "frozen" and "unfrozen", the latter only being considered in the variational process. The quantities Ff and Fu contain parts that vary at each iteration and parts that remain constant. The former consist of matrix elements containing ϕu, the latter consist of matrix elements containing ϕf. In this way, the problem of mxN electrons (where N is a number between 1 and 16 for most biological problems and m is as large as 200) is transformed in a sequence of several problems, each one of tractable dimension. A final phase is to be added by inverting the role of ϕu and ϕf, namely by unfreezing ϕf and by variationally freezing ϕu. The localization of the virtual orbitals retained in the second SCF phase essentially block-diagonalizes the secular equation. Therefore the numerical imposition to the matrix (to be block-diagonalized) corresponds to a physically acceptable approximation. A very brief account of the technique proposed here was previously presented (4).

For special problems, some useful information can be obtained very simply by approximating many matrix elements directly with the point charge approximation.

As an example, let us consider a glycine molecule in the zwitterion form either in vacuum or in the presence of a solvent. We have simulated the solvent by considering 200 water molecules surrounding the zwitterion. The geometry of the solvent molecules was obtained from the last conformation of a Monte Carlo simulation (5). If we are interested in considering the stabilization of the zwitterion due to the solvent, namely if we wish to know the total energy of the zwitterion in the gas phase and in a solvent (at the same geometry), we can simply replace the 200 water molecules by the corresponding net charges (taken as equal to the net charges of a water molecule in the gas phase). The wave function remains localized on the zwitterion simply by not providing basis sets on the water molecules. The results of this computation are as follows: 1) a net stabilization in glycine zwitterion total energy, 2) a corresponding stabilization of the orbital energies and, 3) a variation in the net charges, more specifically a decrease in the electronic population of the hydrogens in the NH_3^+ group, a nearly compensating increase in the electronic charge of the oxygens in the COO^- group and (Table 1) essentially no charge variation in the CH_2 group.

Table 1. Gross charges on $H_3N^+-CH_2-COO^-$ in presence or absence of the field of 200 water molecules.

Atom	without H_2O	with H_2O
H_1	0.600	0.576
H_2	0.600	0.545
H_3	0.562	0.551
N	7.662	7.676
C_4	6.325	6.318
H_1'	0.785	0.779
H_2'	0.785	0.779
C'	5.561	5.561
O'	8.584	8.613
O"	8.529	8.600

We note that the above computation of the system $H_3N^+-CH_2-COO^- + (H_2O)_{200}$ is nearly as fast as the computation of $H_3N^+-CH_2-COO^-$. We recall that the two-electron matrix elements needed to compute the zwitterion are the same in the gas phase simulation as in the cluster simulation. For this reason, the addition of the electrostatic contribution due to the 200 water molecules was performed in a few minutes on an IBM S/370/168.

The second aspect in need of analysis in order to compute chemical systems with 150 to 200 atoms, relies mainly on faster computational techniques for the matrix elements. A contracted gaussian set can be approximated by integrals over the corresponding adjoined gaussian basis set. The adjoined basis set (6) established a one to one correspondence between contracted functions and adjoined functions: the gaussian functions of the adjoined set replaces the gaussian functions of the contracted set for preestablished distances (in the computation of the many-center integrals). The computational gain in speed can be assessed very easily. If the number of contracted gaussian functions in a matrix element $<ab|cd>$ are $n(a)$, $n(b)$, $n(c)$, and $n(d)$, respectively, the number of integrals to be computed is, in general, equal to the product $n(a) \times n(b) \times n(c) \times n(d)$. This number reduces to one for the adjoined basis set. (Typical values of $n(a)$ are between 3 and 5).

However, a new technique to compute matrix elements is becoming more and more practical. In the early thirties, selected integrals were tabulated as functions of parameters; this idea can be re-analyzed. If one restricts oneself to a given type of contracted functions (namely frozen orbital exponents and frozen contraction's expansion coefficients), one can represent the four-center integrals with good accuracy in terms of a very short expansion specific for a given set of contracted functions and valid for any range of distances. The drawback of this approach is that the numerical technique to obtain such expansion is rather laborious; however, the net result is very gratifying since a given four-center integral over a contracted basis set can be computed in a remarkably short time, comparable to the time needed today to compute a simple algorithm like, for example, the computation of an exponential. As a consequence, the present need to store the integrals at each SCF iteration becomes questionable and the problem of transforming integrals from atomic orbitals to molecular orbitals can be fully reconsidered. We are of the opinion that the above technique will drastically change our outlook towards ab-initio computations, and will render obsolete the use of those very small basis set that we have classified as subminimal (for example, STO-3G). This technique for the matrix element evaluation and the previous techniques for the SCF is sufficiently powerful to render a computation on 200 atoms nearly as simple as is today a computation on a ten to fifteen atoms problem. To reach this goal "supercomputers" are not necessary - strictly speaking - but will be most useful, especially because of the expected improved price/performance.

The third aspect that we wish to analyze, concerns the correlation energy problem. Specific aspects, limited, however, to the intermolecular forces, will be considered later in this work. Here we start by noting that most of the standard techniques used today to solve the correlation problem can be considered as modifications and/or simplifications of the full multiconfigurational SCF method. The latter has never been applied, mainly because it appears to be computationally too complex. As a consequence, the configuration interaction method (C.I.) is often used either in the full form, or by including all double excitations and selected triplet and higher excitations or by imposing different constraints based on diagrammatic partitioning of the many body problems. As we have often pointed out, we feel that these techniques are not efficient for large molecular systems, but represent methods for systems with relatively few electrons.

In the mid nineteen-thirties a fundamental paper was presented by E. Wigner (7). The term "electron correlation energy" was first introduced at that time and a basic outline on how to handle the correlation problem was proposed. The basic idea is that in the Hartree-Fock one-electron model one can reasonably take care of the Fermi hole (namely two electrons with the opposite spins cannot come too near). However, in the Hartree-Fock model the electrons with opposite spin are free to come too near to each other and are therefore essentially uncorrelated. As a consequence, one needs to introduce some physical perturbation, not present in the Hartree-Fock hamiltonian in order to create a Coulomb hole. Of course, an embryonal Coulomb hole is presented in the Hartree-Fock method, since otherwise the Hartree-Fock approximation would be physically totally meaningless. The embryonal Coulomb hole is obtained by imposing the shell structure, namely by introducing the quantum numbers. The exchange integrals are the Hartree-Fock way to introduce this embryonal correlation. For this reason, the energy difference between Hartree and Hartree-Fock wave functions is a measure of the "precorrelation" energy (8). The Hartree-Fock model yields an approximate but standardized distribution of the electrons; therefore it can provide basic information on where the need for correlation is most pressing in a given electronic distribution (9). A different way to state the above concept is as follows: the Hartree-Fock function (or its density), is functionally related to the correct electronic distribution, and therefore the electronic correlation energy must be related (through some functional relation) to the Hartree-Fock wave function. A functional form relating the correlation energy to the Hartree-Fock wave function was proposed by Wigner in the above quoted work. These concepts are at the base of well known papers for example by J. C. Slater (10) and Kohn and Sham (11).

If we consider an electron immersed in a sea of electrons, the Coulomb hole will be different depending on the pressure exerted by the remaining electrons on the one considered (alternatively, different terms in the electrostatic potential expansion will be of primary importance in different regions). Therefore electron gases with drastically different densities are associated with different functionals, each one relating the Hartree-Fock function to the electron correlation correction. High density, intermediate density and low density functionals have been proposed (12). These functionals are all expressed in terms of the Hartree-Fock density. The <u>density's</u> selection for the functional, however, constitutes a rather gross physical simplification. Indeed, by introducing electronic correlation corrections, we introduce the equivalent of new trajectories, different from those represented by the Hartree-Fock model. This brings about the need to relate the new trajectories to the kinetic energy of the moving particles. The electron to be correlated can be considered at a fixed position, whereas each one of the other electrons can be assumed to be in motion. Each position is defined at a given time by many different trajectories, each trajectory being weighted with a given probability factor. For example, the many determinants of a configuration interaction wave function can be interpreted as representing the many different trajectories of Hartree-Fock type, variationally weighted (remember the well known correspondence between the momentum space representation and the coordinate space representation). A most interesting attempt to introduce a functional of physically meaningful form is the one presented by Brueckner (13), where density gradient functionals in addition to density functionals have been analyzed (clearly, a density gradient is related to the kinetic energy). It should be noted that the early quantum chemical literature had nearly neglected E. Wigner's proposal, with some exceptions to be found for example in papers by Gombas (12), Clementi (14), and later by Gordon (15). Today the most important problem remaining is the one left open by E. Wigner, namely the definition of the <u>necessary</u> and <u>sufficient</u> conditions relating a given functional to the correlation energy and the derivation of a <u>practical</u> and <u>physically</u> sound functional, based on the above conditions.

To clarify some of the above concepts, let us consider an atomic matrix element of Coulomb type between two electrons. The standard expression is given below:

$$J(k,1) = \phi(k,1) \phi(1,2) (r_{12})^{-1} \phi(k,1) \phi(k,2) dv_1 dv_2 \qquad 1.5$$

and the radial part of this integral is given by

$$\int_0^\infty f(1) \left[\int_0^{r_1} f'(2) dr_2 + \int_{r_1}^\infty f''(2) dr_2 \right] dr_1 \qquad 1.6$$

where $f(1)$, $f'(2)$ and $f''(2)$ are functions of either the electron 1 or 2.

As one can notice, for each position of the electron 1, the electron 2 can coexist in the same volume element however small the latter is selected to be. Equivalently, the trajectory of electron 1 is selected in such a way as to accept a collision with electron 2, at any point along the trajectory. To change the trajectory of electron 2, that is, to create a Coulomb hole, we simply need to impose on electron 2 the condition to never come nearer to electron 1 than some pre-assigned distance. This can be done easily, for example, by introducing a different integration limit for electron 2 as shown below:

$$\int_0^\infty f(1) \left[\int_0^{r_1-\delta} f'(2) dr_2 + \int_{r_1+\delta}^\infty f''(2) dr_2 \right] dr_1 \qquad 1.7$$

where the cut off δ is proportional to the overall electronic distribution, for example, it can be of the form

$$\delta_{pq,rs} = C_1 \left[(\alpha_p + \alpha_q) S_{pq} + (\alpha_r + \alpha_s) S_{rs} \right] * \qquad 1.8$$
$$* \left[S_{pr} + S_{ps} + S_{qr} + S_{ps} \right]$$

where p, q, r, s refer to Slater orbitals of exponent $\alpha_p, \alpha_q, \alpha_r, \alpha_s$, and with overlap S_{ij}.

The analytical solution of the integrals for atomic coulomb and exchange matrix elements with the above referred cut-off have been available since about 1963 (a revised edition of an atomic computer program is available for distribution (16)).

The cut-off value has been chosen as a function of the local density of electrons 1 and 2, thus it is a functional of the density. By

Table 2. Correlation energy in a.u.: best empirical values values obtained with equation 1.9 or with equation 1.7.

Atom		Correct Values	Eq. 1.9	Eq. 1.7
He	^1S	−0.0421	−0.0421	−0.0420
Li	^2S	−0.0453	−0.0476	−0.0523
Be	^1S	−0.0944	−0.0947	−0.0769
B	^2P	−0.1240	−0.1263	−0.1030
C	^3P	−0.1551	−0.1580	−0.1409
C	^1D	−0.1659	−0.1750	
C	^1S	−0.1956	−0.1750	
N	^4S	−0.1861	−0.1882	−0.1897
N	^2D	−0.2032	−0.2090	
N	^2P	−0.2274	−0.2090	
O	^3P	−0.2339	−0.2408	−0.2460
O	^1D	−0.2617	−0.2650	
O	^1S	−0.2985	−0.2649	
F	^2P	−0.3160	−0.2980	−0.3165
Ne	^1S	−0.3810	−0.3594	−0.3980
Na	^2S	−0.3860	−0.3470	−0.4031
Mg	^1S	−0.4280	−0.4417	−0.4371
Al	^1P	−0.4590	−0.4667	−0.4822
Si	^3P	−0.4940	−0.4910	−0.5378

Atom		Correct Values	Eq. 1.9	Eq. 1.7
Si	^1D	−0.5050	−0.5330	
Si	^1S	−0.5200	−0.5320	
P	^4S	−0.5210	−0.5132	−0.5854
P	^2D	−0.5390	−0.5590	
P	^2P	−0.5550	−0.5580	
S	^3P	−0.5950	−0.5818	−0.6380
S	^1D	−0.6060	−0.6310	
S	^1S	−0.6240	−0.6300	
Cl	^2P	−0.6670	−0.6545	−0.7236
Ar	^1S	−0.7320	−0.7310	−0.8324
Li+	^1S	−0.0435	−0.0438	
Be2+	^1S	−0.0443	−0.0447	−0.0428
B3+	^1S	−0.0448	−0.0453	
C4+	^1S	−0.0451	−0.0456	−0.0430
N5+	^1S	−0.0453	−0.0459	
O6+	^1S	−0.0455	−0.0460	−0.0431
F7+	^1S	−0.0456	−0.0462	
Ne8+	^1S	−0.0456	−0.0463	−0.0433
Kr36+	^1S	−0.0470	−0.0471	−0.0433

computing the atomic total energy either with standard coulomb and exchange integrals or with the modified integrals above reported, we have obtained the correlation energies given in Table 2. It is also noted that using wave functions obtained with the modified integrals, some expectation value, like polarizability, improves relative to the standard Hartree-Fock (4) functions.

Years ago (17) we have reanalyzed an alternative method, selecting an Hartree-Fock density functional of the form:

$$E_c = \int 0.02096 \, (1.2 + \rho_m^{1/3})^{-1} \rho^{4/3} dv + \qquad 1.9$$
$$+ \int 0.02096 \, \ln(1 + 2.39 \rho_m^{1/3}) \rho_m dv$$

where ρ_m is related to the Hartree-Fock density by the equation

Table 3. Binding energies (in e.V.) for selected diatomic molecules computed in the H-F approximation, with a density functional, and from experimental data.

Molecule	H-F	Eq (1.8)	Exp.
H_2	3.64	4.69	4.75
Li_2	0.17	1.03	1.14
B_2	0.89	3.06	3.00
C_2	0.79	5.21	6.36
N_2	5.27	9.34	9.91
O_2	1.28	4.64	5.21
F_2	-1.37	1.50	1.66
LiH	1.49	2.36	2.52
BeH	2.18	2.32	(2.2;2.5)
BH	2.78	3.76	3.58
CH	2.47	3.60	3.65
NH	2.10	3.41	3.40
OH	3.03	4.37	4.63
PH	4.38	5.77	6.12

$\rho_m = \Sigma \bar{n}_i \rho_i$ where ρ_i is the density of the i-th natural orbital and n is related to its occupation number by the relation
$\bar{n}_i = n_i \exp[(-2.0+n_i)/2]^2$.

The functional was applied to diatomic molecules, with reasonable success: in Table 3 we present the binding energy computed either in the Hartree-Fock approximation or with the above functional; in addition the results are compared to the experimental dissociation energies.

Table 4. Ionization Potentials and Electron affinities (e.V.) computed and experimental.

Z	Ioniz. Potl. Comp.	Exp.	El. Affinity Comp.	Exp.	Z	Ioniz. Potl. Comp.	Exp.	El. Affinity Comp.	Exp.
3	5.44	5.39	0.75	-	29	8.96	7.72	1.07	1.23
4	9.11	9.32	-	-	30	8.90	9.39	-	-
5	8.63	8.30	0.31	0.28	31	6.19	6.00	0.29	0.30
6	11.51	11.26	1.24	1.27	32	8.01	7.88	1.33	1.21
7	14.74	14.54	-0.43	-0.10	33	10.05	9.81	1.11	0.80
8	13.69	13.61	1.02	1.46	34	9.99	9.76	2.24	2.02
9	17.35	17.42	2.78	3.40	35	12.05	11.84	3.61	3.34
10	21.33	21.56	-	-	36	14.36	14.00	-	-
11	5.49	5.14	1.33	0.55	37	3.82	4.18	0.83	0.49
12	8.31	7.64	-	-	38	5.95	5.69	-	-
13	6.39	5.98	0.51	0.46	39	6.44	6.38	-0.36	0.00
14	8.35	8.14	1.42	1.38	40	6.71	6.84	1.14	0.53
15	10.63	10.50	1.07	0.74	41	5.72	6.88	0.80	1.03
16	10.74	10.36	2.16	2.08	42	5.95	7.10	1.01	1.02
17	13.12	13.01	3.65	3.61	43	7.18	7.28	1.79	0.73
18	15.91	15.75	-	-	44	6.29	7.36	-1.44	-1.13
19	4.25	4.34	1.03	0.50	45	6.78	7.46	0.42	1.23
20	6.46	6.11	-	-	46	7.63	8.33	-0.35	0.63
21	6.84	6.56	-0.70	-	47	6.46	7.57	0.94	1.30
22	7.11	6.83	-0.43	-	48	6.77	8.99	-	-
23	7.14	6.74	-0.11	-	49	5.87	5.78	0.53	0.30
24	5.98	6.76	1.10	0.66	50	7.53	7.34	1.35	1.25
25	7.78	7.43	-2.57	-	51	9.21	8.64	1.31	1.05
26	7.85	7.87	-1.03	-	52	9.14	9.01	2.12	1.97
27	8.41	7.86	-0.76	-	53	10.91	10.45	3.44	3.06
28	8.32	7.63	-0.45	1.15	54	12.32	12.13	-	-

The same type of functional has also been used to obtain the ionization potentials and the electron affinities for several atoms. In Table 4 the computational results are summarized. The characteristic behavior of the ionization potential within each period of the atomic table results from the combined characteristic behaviors of the Hartree-Fock energy and of the correlation correction. In the Hartree-Fock energy, the specific combination of the coulomb and exchange integrals for different states is responsible for the main features of the ionization energy vs. Z value. Such behavior is provided quanti-

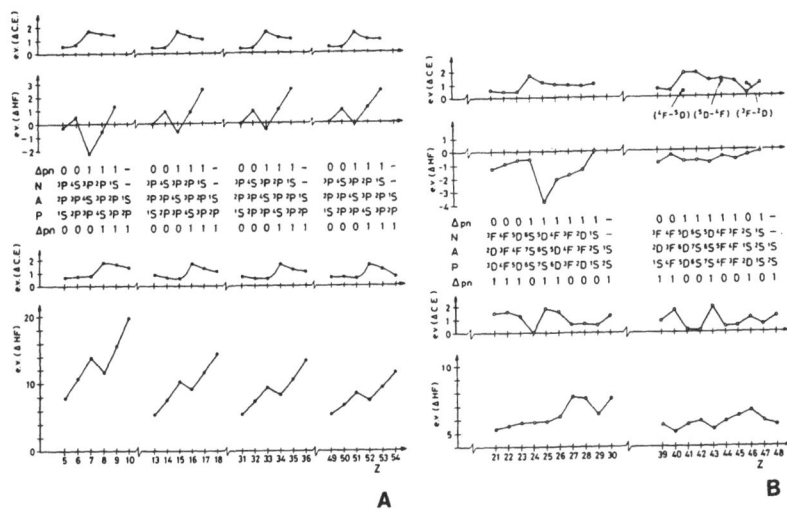

Figure 2. Decomposition of the Electron Affinities and Ionization Potentials in Hartree-Fock energy (ΔHF) and Correlation energy ($\Delta C.E.$) for np electrons (insert A) and for nd electrons (insert B). The first two diagrams (top) refers to the electron affinities; the process atom $\to$ negative ion (A$\to$N) brings about a variation in the pairs number (Δp). The two bottom diagrams refers to the ionization potentials; the process atoms $\to$ positive ion (A$\to$P) brings about a variation in the number of pairs (Δp).

tatively at the bottom inserts of Figure 2 for np^m and nd^m configurations, respectively. The correlation energy correction follows the specific pairing pattern as pointed out years ago (8). The pattern is obtained by using the functional equation (1.9) and is given at the top inserts of Figure 2 for the np^m and the nd^m configurations, respectively. Finally, in Figure 3, the computed and the experimental data are compared. Additional details are, in part, published (18).

From these examples (coupled with the defacto "partial-failure" of the C.I. techniques in macromolecular studies), it should be clear why for many years we have felt, that Wigner's proposal represents a key to obtain the correlation correction in macromolecules. Expected progress in computer architecture and technology (the "supercomputers") will clearly help the C.I. method towards larger molecules, especially if "vector" oriented algorithms will replace presently available "scalar" oriented algorithms. However, even a hundred fold increase in computational speed will soon meet with the "reality" of n^4 or n^5 dependency.

Before concluding this short commentary on the correlation energy problem, we add four comments.

The first one deals with the computation of a functional once the Hartree-Fock function is available. The value of the correlation energy can be obtained by a numerical integration (see for example, 18). To introduce substantial gains in the computational time, the functional can be pretabulated for a given set of contracted functions. Further, the integration volume can be easily decomposed into subvolumes, following the technique proposed (for a totally different purpose) by S. Antoci (19); this partitioning allows the use of a much denser grid near the nuclei than in other regions, a requirement needed

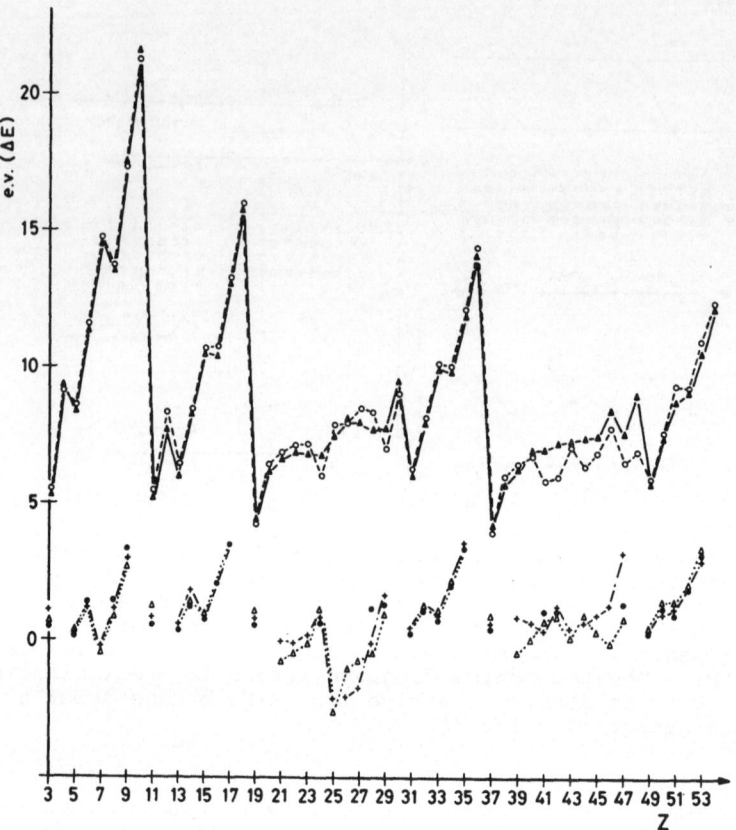

Figure 3. Computed (full triangles) and Experimental (open circles) Ionization Potentials (top diagrams) and Electron Affinities (bottom diagrams). For the electron affinities the "recommended" values, corresponds to the experimental values (full dots). Open triangles correspond to computations with the functional. Crosses correspond to computation with a different method.

to obtain numerical accuracy and to ensure that there exists one integration point at each nuclear position. In addition, for macromolecules Antoci's type integration allows selection of a limited region in the molecule, where one feels that the correlation correction variations are of particular interest. For example, in the study of chemical reactions, one can neglect those regions where one assumes a constant correlation correction.

A second comment concerns the use of the Hartree-Fock wave-function as the selected function for the functional. As known, the term "Hartree-Fock" is often used as a short notation for "reference state function"; the Hartree-Fock model is notably poor in the case of degeneracy or near degeneracy (8, 17). Therefore, a more appropriate definition of the density functional should not refer to the Hartree-Fock model but rather to the reference wave function, built either by a single determinant, or by several. In the diatomic examples given in Table 3, reference wave functions, rather than Hartree-Fock wave functions, have been considered.

The third comment concerns the relationships between the relativistic and the correlation energy correction. As we have pointed out long ago (8), the two are not independent. A functional can operate either on relativistic or nonrelativistic Hartree-Fock densities, thus offering a solution to the correlation problem that will not require that type of expected re-analysis of the problem clearly needed for high Z atoms, but often neglected by many-body theoreticians. As known, (20), the use of configuration interaction and the neglect of relativistic effects might be questionable even for relatively low Z atoms; the total correlation correction is smaller than the total relativistic correction starting at about Z=12. One of the physical reasons for the need to include the relativistic correction in correlation energy studies is that most of the CI type techniques are based on coupling algorithms connecting electronic configurations of specific quantum numbers. Unfortunately, the quantum number scheme often adopted is invalid for electrons in high Z atoms and, likely, marginally valid for inner shell electrons, even for low Z atoms.

Finally, we have noted that recent quantum chemical computations (performed in the last 4 to 6 years) point out a revival of interest in the functional techniques. Some studies starts from Hartree-Fock functions, builds from it the first order density matrices, and re-compute approximately the Hartree-Fock energy making use of functionals. In my opinion these studies are interesting, but possibly fail to grasp the main problem, namely, how to proceed from the Hartree-Fock approximation to the correlation energy correction. Other studies starts with Hartree-Fock wave functions, builds first order density matrices and attempt, via functionals, to obtain both the Hartree-Fock energy and the correlation energy. In my opinion these studies are very interesting, however, perhaps, fail to take full advantage of the fact that the Hartree-Fock energy is the main part of the total energy and the functionals should be considered as a perturbational correction.

1.4 A GENERAL METHOD FOR SIMULATIONS OF A COMPLEX CHEMICAL SYSTEM

The most basic selection one makes to describe a system with some given model, relates to the choice of the adopted statistic; this choice brings about a specific definition of the nature of the "objects" (particles or medium) and concomitant model-equations. "Natural objects" for chemical systems are either nuclei and electrons or atoms and molecules or some continuous medium (rigid or nonrigid). The corresponding statistics are either the Fermi-Dirac statistic or the classical statistic (inclusive of Boltzmann distribution); the corresponding equations are either the Schroedinger equation or Newton's equations (adopted for discrete particles or for continuous medium). In the latter case, of particular chemical interest are aspects dealing with the thermodynamic of reversible and irreversible systems, problems of linearity and nonlinearity, single or multiple solution for stationary states, etc.

Today, quantum chemistry has sufficiently evolved so as to allow us to consider realistically complex systems as an object for numerical simulations. As previously pointed out, however, it is basic to reject the indiscriminate use of quantum chemistry (intended as approximate solutions of the Schroedinger equation) as the only tool to be used for a complex system's description. Such indiscriminate use brings about necessarily gross over-simplifications that could be avoided by using quantum-chemistry as a first step of a many-step methodology aimed at realistic simulations of complex chemical systems.

We briefly summarize an operational procedure to simulate complex chemical systems. Most of the theoretical foundations needed in our approach were known long ago (21). However, little has been done to <u>operationally</u> link different methods that are traditionally kept independent one from the other, despite the fact they represent the successive step of our approach or of any physically reasonable approach. Details on the proposed method will be given in later chapters; here we wish only to present a very brief account of what will follow. Let us start by considering some appropriate subsystem of our complex system. The simplest and most immediate subsystems are the individual, separated molecules. As previously indicated, the size of a molecule for which "decent" quantum chemical simulations are feasible is constantly increasing, either because of new methods or because of the increased performance of computers. "Decent" simulations are those that make use of adequate basis sets and that include electronic correlation corrections, when needed.

The general method is presented as a five-step technique; the output of step (i-1) constitutes the input to step (i). This constraint is basic to develop an operational procedure.

Step 1:   Quantum Chemistry

<u>Ab initio</u> quantum chemical simulations are performed either on one or few molecules. If the complex system is composed of only one macromolecule, then we shall use methods described in the second chapter. Alternatively, the complex chemical system might be composed of many molecules. If ab initio computations are performed in order to obtain inter molecular interactions, then we proceed at first by obtaining the two-body interactions. If three-body interactions are required, then the same procedure used to obtain the two-body correction is followed. Details are given in the fourth chapter. The introduction of the reaction field is assumed under the heading of "quantum chemistry", not because it is a generally accepted procedure, but because it is a necessary interaction in most problems dealing with complex systems.

Step 2:   Construction of Interaction Potentials

From the numerical potentials of Step 1, one obtains analytical potentials. This second step constitutes a most critical aspect to operationally connect quantum chemistry to statistical mechanics. The potentials must be constrained in such a way as to be of an analytically simple form, fast for computational use, transferable from molecule to molecule, standardized both in the form and reliability and, finally, amenable to gradual refinements and extensions. By design we have neglected the possibility to use experimental data as the starting parameters to obtain the potentials. One reason is that often there are not sufficient experimental data available; a second reason is that it is often arbitrary to extract two- and three-body contributions from an experiment obtained at conditions corresponding to a full n-body potential.

Basic to this step is an analysis of how an atom can be characterized when in a molecule. From the valance concepts of Lewis and Langmuir, the valence bond approximation was derived by Heitler, London, Slater and Pauling; an important concept in that language is the valence state concept. Alternatively, from the Lewis-Langmuir concepts, we can arrive at the Mulliken-Hartree-Fock Molecular Orbitals approximation. Heuristically, an important concept connected to the MO model is that partitioning of the electronic density known as the electron population analysis. From the valence bond approximation, we have translated into the MO theory the equivalent of the concept of Valence State and called it "molecular orbital valence state", MOVS. An atom in a molecule can be characterized by its value of the MOVS energy and by the value of its net charge: hybridization, charge transfer and nearest neighbors are included in this characterization. By definition, we characterize an atom of a given atomic number and in a specific electronic environment with a label referred to as "class" for that atomic specification. In this way, atoms of the same atomic number and in the same molecular environment belong to the same class. Hence, by construction, the identification of a "class" for an atom is transferable from molecule to molecule, as long as the atom has the same Z value and class label. As a corollary, atom-atom pair potentials are transferrable from molecule to molecule. Historically, we can note the following evolution in the characterization of atoms in molecules: the first and gross characterization is the one corresponding to the atomic number, (for example, for the carbon atoms, Z=6); then the hybridization characterization (for example, sp, sp^2 and sp^3) and the valence state concepts; now the "class" characterization is added, providing a finer characterization. For example, the carbon atoms in CH_4, CH_3-CH_3, CH_3-NH_2, CH_3-OH, CH_3-SH, have all the same valence, but each one of the carbon atoms belongs to a different class, as clear from the different values of the two parameters, namely the MOVS energy and the net charge value.

The intermolecular interaction potentials must be limited to the two-body Hartree-Fock equivalent interaction, but must include dispersion corrections at the two-body level and induction corrections at the many body level. As later explained in detail, such corrections can be obtained both easily and accurately for intermolecular interactions.

Step 3: Static Properties

The availability of atom-atom pair potential allows us to easily pass to statistical thermodynamics. Static properties are first analyzed. A basic tool in this step is the Monte Carlo, MC, method (22) that allows us to introduce temperature averaging in the complex system. We generally use the MC method at constant volume with periodic boundary conditions and at constant pressure for clusters studies. The introduction of temperature eliminates for example, the unphysical (but currently used) approximation to study solutions at zero temperature. As known, the reactivity of a system is related to its free energy and not only to the internal energy; in this regard we note that the simulation of entropy is now becoming more and more feasible using MC techniques.

Step 4: Dynamical Properties

In this step the time parameter is introduced, for example, in the standard form of molecular dynamics MD (23). Prerequisite for this step is the availability of pair-potentials from Step 1. As an initial condition to describe the system, we use the final configuration obtained from Step 3. Work is in progress on this step in our group. We note, however, that from the current and past literature on applications of MD one could easily obtain a realistic estimate on the importance of this step. The transport aspect is one among several that can be simulated from the time parameter introduction.

Step 5: Continuoum Representation

In this step, the basic coefficients needed to solve, for example, the diffusion equations of a flow are obtained from Steps 3 and 4. As in the case of Step 4, we are only beginning work at this problem, but the very ample literature in fluido-dynamics coupled with the ample literature on biological dissipative systems, and time or/and space fluctuations (24) should be enough to let one understand how much rewarding it will be to unify the scope of Step 5 with the outputs of Steps 3 and 4. Traditionally, this step is the less <u>operationally</u> connected to quantum mechanics and quantum chemistry.

In our opinion, the theory and simulations of complex chemical systems have suffered because of the lack of a general framework, as for example, the one above outlined. Essentially, as with physico-chemical experiments, one can analyze either simple or complex systems without artificial limitations (like the problem of having too many electrons, etc.) so we wish to reach the same type of operative freedom in theoretical simulations. We wish to solve for Newton's equations of complex systems at different levels of resolution. If the particles of the system are electrons and point-charge nuclei (high resolution level) than we impose on Newton's equations the constraint of the quantum numbers (forced on by the Fermi-Dirac statistic). If there are no "particles" in the system, but a continuous distribution of matter (low resolution level), then we impose on Newton's equations the Rayleigh numbers (or its equivalent) as constraints, representing boundary conditions and conservations laws. When we still talk of discreate distribution of particles, but these are atoms and molecules rather than electrons and nuclei, then we are at the intermediate resolution level. Different aspects of a given chemical problem require different levels of resolution hence different statistics and, therefore, different constraints to the equations of motion.

2.0 COMPLEXITY BECAUSE OF THE "SIZE" OF THE LARGEST MOLECULE IN THE SYSTEM

2.1 COMMENTS ON CONFORMATIONAL ANALYSES FOR A SINGLE MOLECULE

Previously we have defined chemical complexity either by considering the number of molecules in the systems or by considering the size of the largest molecule of the system. To the two situations there is correspondingly two different approaches in performing simulations.

In this section we shall briefly review some of the main trends in the study of conformations (conformational analyses). The realization that conformational energy changes in small molecules can be described with relatively simple potential functions, (provided one experimentally knows the height of the barrier due to internal rotation) dates from K.S. Pitzer studies[25,26] in the late 1930. From this starting point, a number of workers, for example, Lifson[27], Liquori[28], Gotlieb[29] and later Pitsyn[30], Nagai[31], Mark[32] and Flory[33] have proposed simple analytical potentials, using experimental barriers obtained from measurements on model compounds. In the mid-sixties it was proven that rotational barriers can be obtained with reasonable accuracy from ab initio computations, often without the need of including the electron correlation corrections. A standard test case was the C_2H_6 molecule and we recall, for example, few early papers by Pitzer and Lipscomb[34], by Clementi and Davies[35], Fink and Allen[36], Pederson and Morokuma[37] and Veillard[38]. On the basis of these computations, Clementi[39] did propose that analytical potentials could be constructed from ab initio computations (on model compounds) rather than from experimental data, using potentials of a form similar to the one tested for empirical potentials. The formalism of the bond-energy analyses (see later, in the next Chapter) that is based on the assumption of the existence of a transferable force field, (characteristic of a functional group) was indicated as a specific technique to obtain analytical potentials; numerical examples have been provided, for example, in studies on the sugar-phosphate-sugar complexes[40,41]. Physically similar, but quite different in the techniques and scope, is the proposal underlying the Force Field Method[42] that also emerged within the last ten years, mainly as a result of the Lifson's school.

In the early seventies a different approach was taken by several researchers, for example, J.P. Malrieu[43]. Since the number of the matrix elements needed in the solution of the Hartree-Fock equations was considered too large to be handled and since early semi-empirical methods, where many matrix elements are ignored (like in the CNDO method), did too often incorrectly yield the conformational stability even in very simple molecules, and new way to parameterize the matrix elements was introduced, heavily based on localization assumptions[43]. Unfortunately, the proposed technique (PCILO) cannot cope with macromolecules, since even with the localization assumption the number of matrix elements is much too large and the number of mutually dependent variables being very large (the rotational angles), the computations must be repeated proportionally as many times, requiring unacceptably long computations.

Conformational studies for complex systems represent, therefore an example of a problem where, likely, quantum chemists have not sufficiently considered that the model to be selected is one where the particles are not nuclei and electrons but atoms or "groups of atoms." As result conformational studies are still often carried out without introducing temperature distribution analyses on the many possible and nearly degenerate solutions; in addition the field effects of the neighboring atoms, an essential factor in conformational determinations, is very often ignored. As a second consequence, it is not surprising that protein conformational refinements have neglected the above quantum-mechanical attempts and are performed with energy optimization methods based on analytical potentials, often obtained from the experimental calibration(44), thus returning to the main stream of progress in conformational studies. Recently, we have concluded a rather extended tabulation of conformational potentials that was carried out during the last five years. The tabulation is to be used in the theoretical determinations of the tertiary structure of proteins, starting from the experimental knowledge of the amino-acid sequence. Such potentials are obtained from computations of the intermolecular interactions between the amino-acids pairs and are transferable to complex systems. With such potentials, the perturbation of the external field can easily be added to standard conformational studies(45).

In conclusion, the main avenue for conformational studies seems to rely on analytical potentials obtained either from experiments or from ab initio computations. The use of direct quantum-chemical methods is feasible only for small systems (model compounds) and it is computationally more expensive.

2.2 A NEW METHOD FOR PROTEIN-SUBSTRATE INTERACTION SIMULATIONS

In this section we consider a different type of conformational problem, namely the one related to conformational variations both in a protein and in its substrate during an enzymatic reaction. This being a fundamental problem in biochemistry, we shall present a rather detailed discussion and a numerical example(46).

It should be rather obvious that a quantum-mechanical computation is not feasible for a system composed of the whole enzyme (several thousand atoms) and its substrate. Even semiempirical methods would be prohibitively expensive in terms of computer time. In addition, it is reasonable to assume, at least as a first approximation, that in an enzyme-substrate complex the chemical reaction represents a rather localized event, and therefore the system to be simulated by quantum-mechanical computations need not be as large as the whole protein-substrate complex. On the other hand, the structural deformations that are induced both on the enzyme and on the substrate, first during the approach of the substrate to the protein, then during the reaction, and finally when the substrate's products leave the enzyme, can be accounted for only by considering the entire system, protein and substrate.

These seemingly contradictory requirements, namely, the need to allow structural deformations in the whole system and the impossibility to quantum-chemically consider the same system for the electronic aspects of the chemical reaction, can be resolved by introducing a net distinction between macrodeformations and microdeformations and by finding appropriate techniques capable of dealing with either one type of deformations or the other. The macrodeformations concern the gross effects primarily related to the conformational changes concomitant to the approach of the substrate to the enzyme and its departure as reaction products. The microdeformations refer to the motions of relatively few atoms that one can assume to take place after the macrodeformation event, though it should be clear that in a real system macro and microdeformations do take place at the same time. The reaction field is included in the computational stage dealing with the microdeformations. The time scale of the two types of deformations is sufficiently different as to justify the separation of motions on dynamical grounds.

The macrodeformations acknowledge that in an enzymatic reaction the interactions between the entire enzyme and the entire substrate must be considered and, to a first approximation, the system stability results from the balance of nonbonded interactions. Therefore, the macrodeformations can be analyzed with those methods that essentially are limited to consideration of energy variations arising from nonbonded interactions, rotational barriers, etc. Computer programs using such methods are in a rather advanced developmental stage and are commonly employed to refine the atomic coordinates of proteins and enzymes obtained from X-ray crystallography. The main features of one of such program called "Refine" (44) are here summarized. "Refine" is designed to deal with large systems such as proteins; starting from a given structural model, it produces a new structure of (approximately) minimum energy. In order to allow complete flexibility to the molecule, the program works in Cartesian coordinates, so that the total molecular energy includes stretching, bending, and torsional terms as well as nonbonded interactions. A simple minimization procedure, which moves one atom at a time for many cycles, is utilized.

The <u>microdeformations</u> acknowledge that in an enzymatic reaction, when the overall structure of the enzyme-substrate complex is approximately settled (by minimizing the energy macrodeformations), local displacements occur that are driven by quantum-mechanical variations of the electronic structure and its energy; these microdeformations constitute the chemical reaction path as usually intended for reactions between relatively small molecules. Here the method of computation employed is the quantum-mechanical one, and the enzyme-substrate complex is truncated to those amino acid residues that physically enclose the atoms that are supposed to react, thus including most of the static reaction field. For this part of our study we have used the IBMOL program(3) that allows one to obtain molecular wave functions and is especially designed for large chemical systems (presently up to about 120 atoms). The truncation represents a delicate point, since neglect of the reaction field brings totally meaningless results.

The enzymatic active site is a very special reaction vessel that not only sterically allows the reagents to come into contact, but also affects the kinetic of the reaction by exerting a very specific field (the reaction field) on the reactants. The sterical constraints and the reaction field are two dominant aspects of an enzymatic reaction and, in general, of a catalytic reaction. Our man-made laboratory vessels are mainly limited to provide a rudimental (because unspecific) sterical confinement but the catalytic "vessels" provide a sterical confinement that is dynamical and optimizes its shape during the reaction mainly by field variations which lower the reaction barriers for a very specific reaction, thus increasing the reaction rates and requiring only a minimum of energy, of the order of KT. Such energy is taken from the whole system (enzyme, substrates, solvent, etc.) and brings about that balance that supports thermodynamically unstable structures, the dissipative structure of the living organisms. Our methodological approach uses, in repeated sequences, first REFINE and then IBMOL, yielding a succession of steps that attempts to model the reaction path in a complex chemical system. Occasionally, within a certain step, the use of the two programs may also be combined in another way. While the program REFINE usually expresses nonbonded interactions by means of empirical interatomic potentials, there may be special interactions for which either data are not available or ab initio potentials appear more reliable. In such cases once a macrodeformation has been performed by adopting a tentative potential, the structure obtained is used as a starting point to derive a potential curve from quantum-mechanical calculations performed on an appropriate subsystem. The ab initio potential curve is then parameterized for use in the program REFINE, and the macrodeformation is repeated. Recently, we have started to replace all the empirical potentials originally present in REFINE, with ab initio potentials; the nonbonded interactions have been the object of particular attention(45).

As an illustration of our approach, we shall consider the reaction of
hydrolysis of a peptide substrate (proteolysis) catalyzed by papain(46).

Papain is a cysteine proteinase, which owes its catalytic action to the
presence in its active site of the essential residues Cys-25 and His-159.
The three-dimensional structure of papain has been determined by X-ray
diffraction by Drenth and co-workers(47), who have also proposed a
mechanism for the action of papain on the basis of the observed binding
of chloromethyl ketone substrate analogs to the enzyme(48,49).

The mechanism proposed by these authors may be summarized as follows.

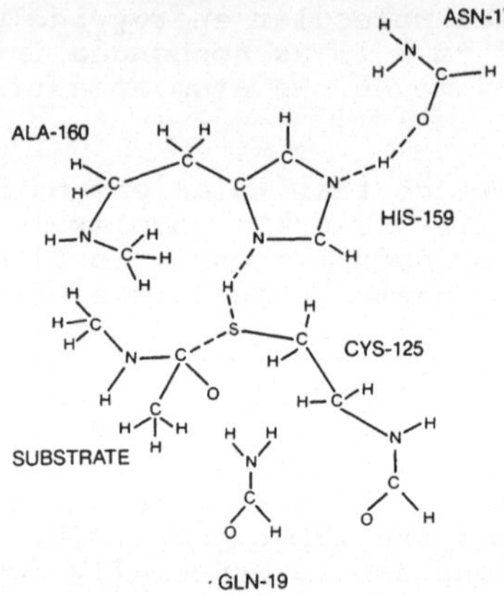

Figure 4. Schematic representation of the essential residues of the
substrate-papain complex as truncated in the ab-initio computations.

In the active site of the enzyme the side chains of the two essential
residues form an ion pair, $ImH^+...S^-$, with the sulfur atom approximately
lying coplanar with the imidazolium ring of His-149 (see Figure 4); the
formation of the ion pair should be favored by hydrogen-bond formation
between the imidazole ring and the side chain of Asn-175. When a peptide substrate binds (noncovalently) to the enzyme, the C=0 group of the
scissile peptide link forms a hydrogen bond, and favors the nucleophilic
attach of the sulfur anion on the carbon, leading to the formation of a
tetrahedral intermediate. In this structure the NH group of the scissile
peptide bond points to the C=0 of Asp-158, while the lone pair of N is
oriented toward the imidazolium ring, which is rotated by 30° from its
initial position. Then the ring donates its proton to NH, and the peptide bond breaks giving as products the leaving amine and a covalent
complex (acyl-enzyme). The second part of the reaction (decylation)
proceeds in a similar (inverse) manner: a water molecule replaces the
leaving group, donates a proton to the imidazole ring, and forms a
second tetrahedral intermediate; then breaking of the C-S bond yields
the second product (acid) and restores the enzyme in the active state.

As previously discussed, the whole reaction path may be idealized as a
sequence of macrosteps, each one reflecting the need to perform a macrodeformation, that is followed by a number of microdeformations (microsteps). In this study we are not interested in obtaining a very detailed energy surface, but only in examining a few critical points in
order to explain our method. In the following we describe a few
initial macrosteps and the associated microsteps.

2.2.1 MACRODEFORMATIONS

Since the aim of this calculation is to show how to determine the

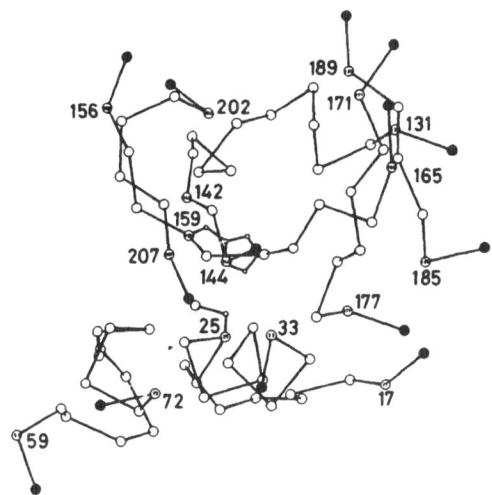

Figure 5. Schematic representation of the portion of the papain molecule considered in the empirical calculations. Only C(α) are shown, besides the Cys-25 and His-159 side chains. Black C(α) were kept fixed.

conformational variations of relatively few residues near the active site, as a first approximation we decided to take into consideration only about one third of the molecule, namely, 73 residues forming and surrounding the active site; this is shown in Figure 5. These residues belong to the following seven segments of the polypeptide chain: Lys-17 to Thr-33; Arg-59 to Leu-72; Ser-131 to Tyr-144; Lys-156 to Gly-165; Ile-171 to Trp-177; Tyr-185 to Ile-189; Leu-202 to Phe-207. In addition, the main chain atoms belonging to the residues adjacent to the seven segments were also included in the calculations, but were kept fixed, as boundaries to the other atoms. We felt that the approximation of considering only a part of papain may affect the residues in contact with the neglected parts of the protein, but much less those residues constituting the active site and essential to the reaction.

At the start of the calculation, we decided to first adjust the side chains of Cys-25 and His-159 (by means of a local energy minimization) keeping the rest of the molecule fixed. Following these preliminary manipulations, all of the 73 residues were allowed to move, thus removing most of the large steric repulsions (notably at residues 18, 20, 132, 136, 138, 202, and 203) and minimizing the electrostatic energy by optimization of several hydrogen bonds and the formation of some new ones. The latter corresponds in some cases to notable displacements of the acceptor or the donor group and to significant changes of the local conformation, relative to the initial set of atomic coordinates.

As a second macrodeformation, we have considered the structure of the papaine subtrate complex. The model substrate N-methyl benzyloxycarbonyl-phenylalanyne   amide (Boc-Phe-Ala-NHCH$_3$) was chosen for this study, since we could start from the X-ray structure of the product of the reaction between Boc-Phe-Ala chloromethyl ketone and papain(48).

The starting coordinates of the -NHCH$_3$ group, corresponding to the amine end of the peptide bond to be broken, were obtained in preliminary calculations; then these coordinates and the experimental coordinates of Boc-Phe-Ala were adjusted to the intermediate structure of the previous step, keeping the protein fixed. Finally, both the substrate and the 73 protein residues were allowed to move.

The microstep associated with this step (see below) is represented by the interaction between model substrate (N-methylacetamide) and a few essential residues of papain, arranged according to the geometry just described.

The next macrostep considers the enzyme and the substrate forming a covalent tetrahedral complex. The purpose of this step was to obtain a low-energy structure of the covalent complex, which is formed upon the attachment of the Cys-25 sulfur anion on the carbonyl of the scissile bond. The system considered is the same as in the previous step, namely, 73 papain residues and Boc-Phe-Ala-NHCH$_3$ as substrate. Preliminary calculations were performed to obtain a first set of coordinates for the complex between -Ala-NHCH$_3$ and Cys-25:

```
      H3C-NH      O-
             \   /
              C
             / \
   -NH-C(α)H   S(γ)-C(β)H2-C(α)H
        |
       CH3
```

In these computations the experimental coordinates of all the surrounding residues were first used as rigid constraints, then the side chains on Gln-19 and His-159 were allowed to adjust. Some assumptions on the geometry of the tetrahedral complex were necessary: the ideal tetrahedral value was used for the bond angles at the C atom, bond lengths C-O = 1.31 A and C-S = 1.86 A were taken, following the computations by Scheiner, et al.(50). This set of coordinates was used in a subsequent energy minimization.

We proceeded as in the previous step: the coordinates of Boc-Phe- were first adjusted to the protein structure. Then the energy of the whole system was minimized.

<u>Additional steps</u> are required to represent the formation of the acyl-enzyme, as the amine product leaves the enzyme. These are not analyzed since our goal is to present the method, rather than a study of a specific enzymatic reaction. Let us now consider the microdeformations associated to the above macrodeformations.

2.2.2 MICRODEFORMATIONS

The problem concerning the stability of the ImH$^+$...S$^-$ ion pair was originally approached by Broer, et al.(51), who considered the system CH$_3$-S...H...Im with a minimal basis set. In their study the field effect was neglected and no macrosteps where considered, namely the computation was performed in the way standardly accepted prior to our work.

In our <u>first microstep</u> the essential residues for the ImH$^+$...S$^-$ system are those that are sufficiently near to the N of His-159 and S of Cys-25 atoms involved in the ion pair formation. In this way we ensure the inclusion of the most important static part of the reaction field. The atomic coordinates were taken directly from the macrostep 1. The system, hereafter referred to as system I, includes the side chain of Cys-25 (represented by C(β)H$_3$-SH), the whole residue His-159 (but with the peptide NH- substituted with a hydrogen) with part of the adjacent residue Ala-160 (represented by $C(\alpha)H_3$-NH-), and finally be the side chain of Asn-175 (approximated by a formamide molecule). The four hydrogens substituting for truncated parts of the chain are positioned on the truncated bonds at a distance of 1.09 A from $C(\alpha)$ or $C(\beta)$ and of 1.10 A from $C(0)$. The microdeformations, here considered involve mainly $H(S)$ (namely, the hydrogen between S and N ($\delta 1$) of His-159). We have considered two positions for the S atom. For the case of CH_3S^- we have computed the C-S bond length using a double-zeta basis set and have obtained a distance of 1.91 A. This distance is used to define one of the S positions. For the CH_3SH system the C-S bond length is the experimental one as given for cysteine(25). We note that the use of the two different bond lengths affects the $S...N(\delta 1)$ intermolecular distance only slightly, since the C-S bond is approximately orthogonal to the line joining the S and $N(\delta 1)$ atoms. In the following we differentiate between system 1A and system 1B, according to whether we use the C-S bond distance for the ionic CH_3S^- system or for the neutral CH_3SH system, respectively. Therefore, system 1A represents the ion-pair form and system 1B, the covalent form. All computations have been performed with the IBMOL computer program.

In Figure 6, insert A, we present the stereoview of system I, with inclusion of several of the computed positions for the $H(S)$ moving along the S-H and N-H directions; note that the two directions form an angle.

For the 1A system we have determined that there is a minimum at the $N(\delta 1)$-$H(S)$ distance of 1.17 A corresponding to the ionic form, and a second shallows minimum at the $N(\delta 1)$-$H(S)$ distance of ~1.8 A (corresponding to a bond length $S(\gamma)$-$H(S)$ of ~1.6A). For the 1B system, we learn that there is a minimum at an $S(\gamma)$-$H(S)$ distance of 1.50A.

Comparing the lowest minimum for system 1A with the minimum for system 1B, we find that the system $ImH^+...S^-$ is less stable than the system Im...HS by 7.5 Kcal/mol (0.0119 a.u.).

Let us analyze the field effect, namely the effect of Asn-175 and Ala-160 on the 1A and 1B systems, by removing first Ala-160 and then Ala-160 and Asn-175. The two subsystems are called II and III, respectively. In Table 5, we report the ab initio energy for the systems I, II, and III, considering only the $H(S)$ positions A3 and B4.

The above computations are performed by keeping $H(\epsilon 2)$ of Asn-175 and $H(N)$ of Ala-160 fixed at the N-H standard distance of 1.00 A. However, such distances are too short, since $H(N)$ of Ala-160 is H bonded to the S atom in Cys-25, and H ($\epsilon 2$) of Asn-175 is H bonded to the oxygen atom of Asn-175. Thus we have repeated the computations for systems I at the $H(S)$ positions corresponding to B1, B4, B5, A2, A3, and A4 but with the two N-H distances of 1.08 A.

The conclusions that can be inferred from the computations of these microdeformations are rather obvious. One <u>cannot compare the enthalpic variation obtained in a small system</u> (for example, System III) <u>with what occurs in the entire enzyme</u>. Alternatively stated, the inclusion of the field effect is essential to ensure a minimum of realism in the simulation.

On the other hand, the entire enzyme does not need to be considered in full detail, but only the essential residues need to be accounted for. An enthalpy study of a local effect, like the ion-pair formation discussed here. The essential residues need not be as many as those needed in the macrodeformations, but should reproduce the electronic environment not only to the first order (for example, Cys-25 and His-159) but also to the second order (essential residues directly connected to those of the

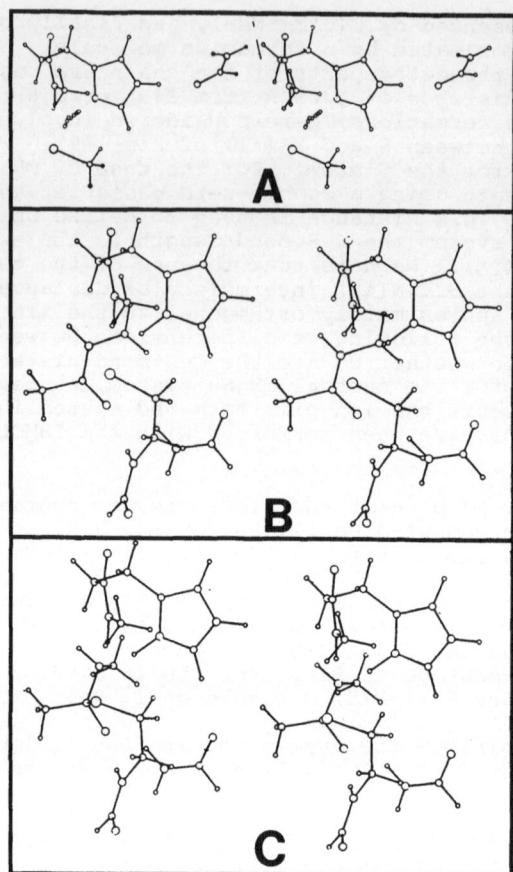

Figure 6. Insert A. Stereoscopic view of system 1 in the conformation used for ab-initio computations. Atom H(S) is shown in the various positions considered along bonds S-H and N(1)-H. Insert B. Stereoscopic view of the non-covalent complex (system 5) as studied in the ab-initio computations. Insert C. Stereoscopic view of the tetrahedral intermediate (system 6) as studied in the ab-initio computations.

Table 5. Stablization effect in the pair ImH+...S- due to the presence of the residues Asn-175 and Ala-160.

System Code Meaning		E(A-B) (Kcal/mole)
III	Cys-25,His-159	18.0
II	Cys-25,His-159,Asn-175	13.4
I	Cys-25,His-159,Asn-175,Ala-160	7.5
I'	Cys-25,His-159,Asn-175,Ala-160	5.3

first order). This in a first order approximation; additional residues can be included as point charges.

The first-order effect for the ImH$^+$...S$^-$ and the Im...HS systems brings about an energy difference of 18 kcal/mol; the second order lowers this difference to about 7.5 kcal/mol. At this stage, third-order effects need to be included. This has only been partially done in our work, but also with this limitation, the energy is lowered from 7.5 to 5.3 kcal/mol. Since the optimization is not complete, here we have computed only an upper limit to the static field approximation. However, even as now given, the energy difference is sufficiently small that proton tunneling can be considered as a real possibility.

The <u>second</u> set of quantum mechanical computations, our <u>microdeformations</u>, concerns the complex formation between papain and the substrate. N-methylmethylacetamide, (NMA), CH_3-CO-NH-CH_3, was chosen to simulate the substrate (see Figure 6, insert B) since it contains the peptide link -CO-NH- in an electronic environment sufficiently similar to that of a peptide chain. The simplified system chosen to represent the essential residues of papain, hereafter called system IV consists of a) His-159 and part of Ala-160 as described in the microstep 1 for b) system I, the whole residue Cys-25, except for a C=O group substituted for by a hydrogen atom, and a formyl group, HCO, representing the end chain of the adjacent residue Ser-24, and c) a formamide to simulate Gln-19.

In system IV we have not included Asn-175, since it is sufficiently far from the chemical region simulated in this microstep. The final geometry is given in Figure 6. The computed ab initio total energies for system IV and NMA are E(IV)=-1318.916711 a.u. and E(NMA)=246.035429 a.u., respectively; the basis set is the same as previously for microstep 1. The computed total energy for system IV interacting with NMA (this constitutes system V) is E(V)=-1564.958259 a.u., therefore the complex interaction energy is ΔE(V)-E(IV)-E(NMA)=0.006119 a.u., namely, the complex represented by system V is stable by 3.8 kcal/mol. Next we ask what is the reaction field effect on the ΔE of the complex. Again, the answer is obtained by neglecting some of the residues and by performing an ab initio computation on the (NMA+CH_3S^-) complex. Since E(CH_3S^-)=-436.258971 a.u. and since the energy of the (NMA+CH_3S^-) system is computed as E(NMA, CH_3S^-)=-682.282546 a.u. we conclude that the (NMA+CH_3S^-) system has an interaction energy of E(NMA,CH_3S^-)-E(NMA)-E(CH_3S^-)=+0.011354 a.u. (7.4 kcal/mol). In conclusion, the presence of the residues His-159, Ala-160, and Gln-19 brings about a stabilization of the complex of 7.7+3.8=11.2 kcal/mol. As in the microstep 1, the ΔE of -11.2 kcal/mol represents only an estimate, since further optimization of geometry is required and this will lower the value of ΔE.

As a final example of our technique, we report about a third microstep, where we have considered the formation of a tetrahedral intermediate for the enzymatic reaction. In Figure 6, insert C, the atoms considered in this microstep (called system VI) are reported, and represent the reaction intermediate. The system considered is composed, as before, of Gln-19, Ser-24, His-159, Ala-160, Cys-25, and NMA. The computed ab initio total energy, is -1564.915250 a.u., to be compared with the value of -1564.958259 a.u. for the complex. The tetrahedral intermediate is less stable by 0.043009 a.u. (or 27.0 kcal/mol).

There are a number of preliminary conclusions to be drawn from this result. First of all, the geometry obtained with REFINE is likely to be insufficiently optimized. Note that the interaction energies for the bond lengths and bond angles of the tetrahedral complex are large and even a very small variation in the geometry will bring about large energy variations. A second possibility to consider is that the mechanism proposed by Drenth, et. al.(47-49) might be somewhat oversimplified.

To partially analyze the above problem, let us start by considering what is the value of ΔE between the complex NMA + CH_3S^-, as considered in microstep 2, and the tetrahedral intermediate (NMA+CH_3S^-); performing an ab initio computation and correcting for the basis set difference truncation error, we obtain a value of 0.061308 a.u., namely, the tetrahedral intermediate is even less stable (by 38.5 kcal/mol) if we do not consider part of the enzyme's environment. The stabilization of about 11 kcal/mol brought about by the enzyme's environment is of the same order as the stabilization obtained in microsteps 1 and 2. Therefore further geometry optimization for the entire system VI will easily lower the barrier. We arrive at the same conclusion by considering the results of another computation, where the tetrahedral intermediate for the (NMA+CH_3S^-) system is computed. The tetrahedral form has a total energy of -682.510867 a.u., hence is 7.0 kcal/mol less stable. Therefore we conclude again that the barrier of 27 kcal/mol is partly due to incorrect input geometry and additional computations are required.

In conclusion we have faced the methodological problem concerning the possibility of studying enzymatic reactions with theoretical and computational methods. From the results presented we are of the opinion that the distinction between macro and microdeformations allows us to study such complex systems. We note that in the specific case of papain, the subsystem representing the active site and the substrate, represents a rather complex case among the presently known enzymes. This observation holds for the computational techniques concerning both macrodeformations and microdeformations. The specific study of the papain activity here reported is still at a preliminary stage. However, one aspect seems to emerge in a clear way: the interactions due to the presence of protein residues <u>not</u> <u>directly</u> involved in a local chemical transformation represent a basic stabilization factor that brings enthalpy variations to the range characteristic of enzyme catalysis, namely to the KT range. Presently, we have undertaken a study aimed at including the field effect in a more economical way: there are two avenues, at least. For residues at intermediate distances from the reaction site, we can use an improved version of the adjoined basis sets(6); for residues at long distances point charges might be sufficient. An alternative way is to introduce dispersion and induction corrections using the algorithms we shall explain in detail in the fourth chapter of these notes.

2.3   FURTHER IMPROVEMENTS FOR ENZYMATIC REACTION SIMULATIONS

The method presented above provides a good starting point for further improvements that are now briefly considered. First, we note that the Refine program can be extended to include the water molecules bound to the enzyme and to the substrate. In this way, the solvent molecules of the first shells, those bound to the macromolecule, can be realistically included in the geometry optimization. The interaction potential to perform such optimization is available (see the discussion in the next chapter). Physically, it does not make too much sense to fully optimize the coordinates of the enzymes and of the substrate, for example, by rotating some bond by a few degrees (and thus gaining a very small amount of energy) and at the same time to neglect to optimize those water molecules that are bound to the enzyme by several Kcal/mole. It is noted in addition, that the energy needed to substain the reaction is provided by the solvent, via kinetic energy transfer, polarization and induction effects and charge-transfer. A second basic improvement is an estimate of the ΔS in the reaction steps, since the kinetic is governed by the free energy variations and not only by the enthalpy variation. This can be done at the macrodeformation level, since after optimization one can perform small displacements and thus obtain the entropic vibrations via vibrational frequencies variations.

A third basic improvement is connected to the realization that the reaction field has both a <u>static</u> and a <u>dynamic</u> component. The static component is essentially the one we have considered by including a larger and larger fragment in the quantum-mechanical simulations (microsteps). The dynamic components are expected to be, in general, rather small with the exception of those vibrational modes that might involve a large fraction of the enzyme and are of a collective nature. It is very tempting to postulate that macromolecules have characteristic vibrational collective modes extending over the macromolecule, with frequencies specific for each enzyme. Such very low frequencies will modulate the static reaction field and propagate into the solvent near the macromolecule, thus providing one of the mechanism of recognition of the enzyme for a particular substrate. (The old analogy between molecular vibrations and electrical networks coupled with the very modern circuit analysis techniques developed to design computer's logic might provide a convenient methodology to study the macro vibrational modes in enzymes and biological macromolecules, and answer the important question on whether macromolecules possess specific vibrational macromodes and thus resemble biological clocks with characteristic time interval units.) The kinetic of the O_2 intake by emoglobine strongly suggest the need of a mechanism related to some of the time dependent aspects here suggested.

In conclusion, we can consider the system participating in the enzymatic reaction as composed of several subsystems: the enzyme, the substrate, the reaction products, the solvent, etc. The enzymatic reaction starts at same time, $t(i)$, and ends at same time $t(f)$; during the reaction time, $\Delta t = t(f) - t(i)$, a number of steps take place that can be mapped either on a reaction path or on a reaction-time scale. Each step is associated with an energy variation distribution, yielding either energy barriers or energy minima. Let us consider the internal energy (enthalpy). By neglecting the static reaction field components, the energy barriers generally are much too high for the energy available in the system; inclusion of the field lowers the barriers to energies of the order of KT, namely transform the reaction from an impossible one to a possible one, even if not necessarily too probable. Let us consider the free energy of the system, by adding $T\Delta S$. The free energy barriers are likely lower than the enthalpy barriers, since the system is in solution, namely in an ideal situation to have a reaction driven not only by enthalpy but also by entropy. The total system is assumed to have a given average energy and the subsystems provide an energy distribution centered around the average energy. Be E the energy of the total system and be $E(i) = E + E(i)$ the energy of the i-th subsystem, (a point of the distribution around E). Energy can be transferred from the i-th to the j-th systems, if $E(i) > E(j)$. If the dynamical component of the reaction field has frequencies of the order of Δt, then the energy transfer could be modulated in such a way as to have energy donation from the solvent to the enzyme for a time interval Δt and energy acceptance from the solvent to the enzyme in the following time interval. If this hypothesis is correct, then the inclusion of the static reaction field components and of the entropy contributions would represent only few necessary conditions to sustain the reaction, whereas the additional inclusion of the field modulation would constitute other necessary conditions. In other words the enzymatic vessel not only changes in time, but the changes are modulated to turn off or on the reaction, thus controlling the energy flow in function of time.

3.0 ANALYSES OF CHEMICAL BONDS

3.1 INTRODUCTION

The analysis of a system composed of two or more interacting subsystems is often pursued, in ab initio studies, by taking as hamiltonian the same hamiltonian used for atomic systems but with the inclusion of more than one nucleus; the nuclei are kept at fixed positions as is done in atomic computations (Born-Oppenheimer approximation). Thus, the system contains as particles, the m nuclei and n electrons; we briefly designate it as the m,n-system. For example, in considering the interaction of NH_3 with HCl, a reaction path was obtained (53) by considering a number of different relative positions for the two subsystems, NH_3 and HCl, and by repeating standard SCF-LCAO-MO computations at each position; however, nowhere in the hamiltonian is there an explicit reference to the decomposition of the two subsystems. In recent years this method has been called "supermolecule approach", a rather unnecessary and inappropriate name, in our opinion, since the term "molecule" has a very precise definition that conflicts with the implications in the term "supermolecule".

It is customary in the study of complexes to distinguish between inter-molecular and intramolecular forces. However, strictly speaking, these definitions do presuppose a definition of the term "molecule". Such a definition which is not contained, as above indicated, in the hamiltonian at least as it is generally used. On the other hand, the traditional definition of "molecule" either derives explicitly from the theory of gases, or is implicitly based on the theories of "chemical bonds". In turn a definition of chemical bond brings us necessarily to a discussion of the chemical formulae. If we consider, even briefly, some of the steps from Mendeleyev (1869) to Van't Hoff (1874), to Lewis (1909), we soon realize that one of the most valuable theoretical concepts in chemistry is the information content implicit in the chemical structural formula. Writing benzene by connecting the six carbon atoms with an hexagon and by connecting each hydrogen to only the nearest carboh, did represent a theoretical achievement of such magnitude as to find relatively few equivalents in today's quantum chemistry. It is interesting to note that in the work of Mulliken, we feel a hesitation: the n electrons share the m nuclei by means of the molecular orbitals, but the charge transfer complex was for a long time described in the terms of concepts nearer to the valence bond approximation and the description of the MO electronic structure was only later simplified and condensed by introducing the concept of gross and net charges (on an hybridized atom) and of overlap electronic population in between two atoms. The need to recast the orbital description in something less delocalized is also at the basis of all the various localization techniques put forward in the last decade. Moreover at the root of such hesitations and needs, there is probably the difficulty to provide a simple, pragmatic and valid description originated from quantum chemistry but as equivalent as possible to the everyday chemical formulae.

In this chapter we shall 1) describe a simple, pragmatic, although arbitrary method (called bond energy analysis) for deriving chemical formulae directly from quantum chemical computations, 2) provide a definition from this formalism of the terms "atom" and/or "molecule", constituting the m, n-system and, as a corollary, a definition of intermolecular forces, 3) use the bond energy analysis as a means of discriminating between different reaction paths, and 4) use the bond energy analysis to obtain simple and transferable atom-atom potentials to describe intermolecular interactions.

3.2 BOND ENERGY ANALYSIS

The name "Bond Energy Analysis" does not imply that we pretend to present a definitive analysis of the theory of bonds in molecules, since we are still attempting to understand the natures of the chemical bonds. It is merely a decomposition of the total energy (analysis) into bonding and nonbonding pairs of interacting atoms (bond energy, for short). In the early stages (54) we found it difficult to limit our analysis only to pairs of atoms, and we included triplets and quartets as well, since these very naturally appear, due to the form of the orbitals (L.C.A.O.) and of the hamiltonian. We then found that we had too much data and it was rather impractical to carry out any simple analysis; in addition we wanted to move closer and closer to every day chemical language. In what follows, we shall briefly recall a few early developments and then move to new one.

<u>Old Formalism and Some Applications</u>. The set of m atoms constituting a molecule can be represented in several ways. The simplest representation is limited only to a statement of the number and kinds (Z value) of atoms constituting the molecule, namely "the gross chemical formula"; thus we write, for example, C_6H_6 to designate benzene. A more refined representation schematically adds, to the gross formula, topological relationships (bonds) connecting elements of the set (structural formula); in this representation among the $m(m - 1)/2$ connections, representing pair-wise interactions, only a few are selected and, precisely, those that connect nearest-neighbors and in such number as to satisfy the "valence" of each one of the m atoms. On each bond <u>two-electrons</u> are allowed with opposite spin. The formal representation is the one proposed by Pauling and Slater, known as valence bond approximation. The "connections", called bonds, can be either single or multiple (single bonds or multiple bonds). The organization of the atoms into the Atomic Periodic Table provided the first theoretical basis for a representation of a molecule, since it allows one to obtain analogies between the valence of the atoms and with Van't Hoff's findings, (1874) the bases were set for the representation of a molecule as a sterical structural formula. The next important step is provided by Lewis "electronic rationalization" of a chemical formula; as known, Lewis introduced the model of an electron-pair, placed between two bonded atoms, and a prequantum mechanical explanation of ionic and covalent bonds. Finally, by formulating the concepts of "lone pair", Lewis introduced a concept that today could be defined as the "latent" bond for a chemical reaction as distinct from the "actual" bond in a chemical formula for an isolated molecule. The Pauling-Slater formulation rationalizes quantum-mechanically aspects of the Lewis theory.

A different quantum-mechanical description of the electronic structure of a molecule followed soon with Mulliken's LCAO-MO model (1935), presented as an approximation to the Hartree-Fock model (in turn proposed as a first step towards more accurate wave functions). Much later (1955), Mulliken attempted to reconcile by means of the electron population analysis formalism (1) the traditional idea that a molecule is composed of a set of m atoms. In the MO underlying assumption, a molecule is composed of a set of m fixed nuclei and n electrons. However, it was difficult to obtain from the wave function a representation as simple and powerful as the one provided by the structural formulae. As known, the chemist needs to associate with a chemical bond both the electronic density and its energy. The statements "forming" and "breaking" a bond imply both density and energy and in a rather subtle way, since the bond energy can be either attractive or repulsive, whereas the density can only be positive.

In what follows, we shall stress the energetic aspect of a chemical bond, rather than the electronic density aspect, even if the two are, clearly, interdependent. The electronic density aspect of a bond was fairly well understood since the very early papers on the MO theory by Mulliken (55) and Herzberg (56). The molecular orbitals were divided into bonding and antibonding, the latter being as basic to the molecular binding as the former. A bonding orbital builds up electronic density in the region in between the two bonded nuclei, whereas an antibonding orbital often substracts electronic density from the same region. Thus H_2 with one bonding orbital and He_2 also with one bonding orbital have drastically different dissociation energies since He_2 has also an antibonding orbital not occupied in H_2. Strictly speaking, the inner shell electrons in a molecule with point symmetry are combined in either bonding or antibonding orbitals. This subdivision assigns a phase and a nodal character to those molecular orbitals that are built up with the inner shell electrons. Physically, the symmetry's constraints for the inner shell electrons do not hurt, but at the same time they are not terribly important unless one is interested in removing inner shell electrons, since the nuclear field on the inner shell electrons is so strong that we can consider the rest of the molecular field as a relatively small perturbation. In conclusion, the bonding and the antibonding nature of the molecular orbitals provides the basic distinction on the electronic density differences between two bonded nuclei and two nonbonded nuclei. Graphical display of the electronic density neatly brings out these features and allows for an immediate visualization of the density aspect of bonds, hence for the density aspect implied in chemical formulae. As known, in writing a formula one solves the problem with a binary decision; between two nuclei there either is a bond or not. Equally well known is the fact that standard chemical formulae and the quantum mechanical density skeleton (that reflects the bonding and antibonding aspect of the molecular orbitals forming the total wave function) are in overall agreement; more importantly, from a good wave function the computed density is in agreement with accurate X-rays scattering intensities and with the basic aspects of the old valence characterization of an atom in a molecule.

Mulliken's electron population analysis is an early attempt to provide a condensed, simple and quantitative picture on the density variation between any two nuclei of a molecule. Its extreme simplicity, the analysis was proposed when computers were hardly used, cost the inclusion of several assumptions that are at the origin of an extensive (and at times marginally useful) series of papers attempting to improve the original assumptions. Of interest, however, are the rather recent techniques proposed, for example, by Bader (57) and Hall (58); Bader provides a beautiful visualization of the electronic distribution but fell short in distinguishing single from multiple bonds, an important characteristic of the chemical formulae and often cannot distinguish between a binding and a repulsive situation, like in H_2 and He_2.

Let us now turn to the energetic aspect of a chemical bond in polyatomic molecules, the subject of the bond energy analysis approximations.

In this section, the early concepts (54) are summarized. We start by recalling that the molecular orbital $\phi(i)$ can be written as a linear combination of a finite set of functions that are centered on the nuclei (or, if not centered on the nuclei, can always be expanded into an equivalent set with the origin at the nuclei). The bare nuclei hamiltonian matrix elements $<\phi(i)|h_0|\phi(i)>$ are expressed as matrix elements over the basis set positioned at three nuclear positions at most; the two-electron operators of the hamiltonian, $1/r(1,2)$, is equivalently characterized by matrix elements at most at four nuclear positions (we shall use the word "center" as a short expression for the expression "nuclear position, where a basis set is centered"). If we designate the m nuclei with the indices A, B, C and D, then the total energy of the molecule can be written as follows:

$$E = \sum_{A} E(A) + \sum_{A \neq B} E(A,B) + \sum_{A \neq B \neq C} E(A,B,C) + \sum_{A \neq B \neq C \neq D} E(A,B,C,D) \qquad 3.1$$

where the last summation drops out for triatomic molecules and the last two summations drop out for diatomic molecules. Thus, the total energy for any molecule can be written in general as the sum of one-center, two-center, three-center, and four-center energies; with obvious notation we write:

$$E = E_1 + E_2 + E_3 + E_4 \qquad 3.2$$

The terms in E_1 constitute the sum of the atomic energies of the atoms when in the molecule. Clearly, each term $E(A)$, $E(B)$,... differs, if compared to the energy of the atom in the ground state. Generally, $E(A)$ is higher (less bound) than the corresponding energy of the separated atom; however, for strongly electronegative atoms, the opposite is true. The terms in E_2 contain both the nuclear-nuclear repulsion (positive energy term) as well as all those matrix elements where two nuclei appear as indices. The terms $E(A,B)$, $E(A,C)$, $E(B,C)$,... of E_2 are much smaller than the one-center terms; the four-center terms in E_4 are due only to the electron-electron repulsions. The three-center terms in E_3 are smaller than the two-center terms but larger than the four-center terms.

3.3 ONE-CENTER ENERGIES AND THE MOLECULAR ORBITAL VALENCE STATE

Let us consider in more detail the one-center bond energy term. Since the early thirties, there have been attempts to describe some aspects of the electronic structure of molecules by making use of the concept of the valency state (59-62). The concept of valency state is most naturally obtained in the valency bond approximation (60-62) but it has also been defined in the molecular orbital approximation (56,63). The valency state concept postulates a fictitious atomic state (with no regard to maintaining L and S as good quantum numbers) which reproduces as closely as possible the electronic distribution of an atom when in a molecule. Thus, the valency state concept is, in its origin, connected with the assumption that there "are atoms in molecules", or at least, that we can identify atomic substructures in a molecular electronic structure.

We have distinguished between the valency state of an atom at infinite separation from the remaining atoms of a molecule, and the valency state of an atom at finite (e.g., near equilibrium) separation from the remaining atoms of molecule. We refer to the former type as valency state standard (VSS), and to the latter as molecular orbital valency state (MOVS) (63). The VSS differs only in the energy, not in the density; namely, we take the density as given at a finite internuclear separation and we compute its energy either in presence or in absence of the remaining atoms of the molecule.

More exactly, we define the VSS as the one which can be obtained in the Hartree-Fock formalism by relaxing the constraint that L and S are good quantum numbers and by making appropriate mixtures of pure states (63), and by imposing that it keeps some predetermined density determined by the molecular wave-function.

We define the energy of the MOVS as the one-center energy obtained from the bond-energy analysis. For a given atom the MOVS will vary in energy from molecule to molecule, depending on the neighbors and on the molecular geometry. We realize, however, that our definition is an arbitrary one, since any decomposition in partial energies (of the total energy in a system of interacting particles) is arbitrary. We also note that the atomic density of a MOVS is transferable from molecule to molecule, thus providing the obvious starting point for obtaining nearly SCF wave functions for even very complex systems at a minimum of computational storage cost. Indeed there is no need to perform the standard SCF analysis, if one knows the "atom in the molecule" densities. Such functions are definite as MOVS-LCAO wave functions, and we shall discuss the corresponding formalism elsewhere (64).

3.4 TWO-CENTER BOND ENERGY: BENZENE

A few years ago (65) the ground-state energy and wave function for benzene (at the experimental geometry) were computed in the Hartree-Fock approximation using an extended basis set of gaussian functions (66) (no polarization functions were included in this basis set). The total computed energy is -230.67187 a.u., about 0.15 a.u., from the estimated Hartree-Fock limit (67).

We have partitioned the total energy making use of the bond energy analysis formalism above outlined.

The one-center energy for each hydrogen is -0.1299 a.u., and for each carbon -36.4295 a.u., that is, 0.3201 and 1.2591 a.u. have been lost by each hydrogen and each carbon atom, respectively from their ground states in order to reach the molecular orbital valency state. The two-center contributions are given in Table 6.

Table 6. BES Energies for Benzene (a.u.)

No.	BES Type	Energy (a.u.)	Percent
1	C_1—C_2	-7.7664	42.3
2	H_1—C_1	-4.5534	24.8
3	H_1—C_2	+2.1024	11.4
4	C_1—C_4	+1.2315	6.7
5	C_1—C_3	-0.7866	4.3
6	H_1—C_3	+0.7296	4.0
7	H_1—C_4	+0.4350	2.4
8	H_1—H_2	+0.4122	2.2
9	H_1—H_3	+0.2634	1.4
10	H_1—H_4	+0.1107	0.6
Total		18.3912 (*)	100%

(*) Absolute Value.

The three-and four-center energies are all smaller than 10^{-3} a.u. (and cluster around 10^{-4} and 10^{-5} a.u.); such terms therefore, will not be discussed, since we are interested in the main picture that emerges from the bond energy analysis for the benzene molecule. We note, in addition, that in general, the three-center terms appear as positive (repulsive) whereas the four-center ones are negative (attractive).

The following observation can be made, limiting ourself only to the two-center bond energies, E_2. Let us first consider the carbon atoms designated as C_1 to C_6 (located one after the other around the ring). There are three possible types of interactions between the atoms: nearest neighbor carbon atoms, next-to-nearest neighbor carbon atoms, and opposing carbon atom (examplified by the C_1--C_2, C_1--C_3 and C_1--C_4 interactions). These types of interactions will be hereafter referred to as "bond energy structures" (BES) in analogy to the valence structures. The first two BES are attractive (negative sign), the third is repulsive (positive sign); the corresponding energies are -7.7664 a.u., -0.7866 a.u. and $+1.2315$ a.u.; summing (absolute value) all three of them (9.7845 a.u.), the percent contributions of these BESs are 79%, 8%, and 13%, respectively. (The total is 18.3912 a.u., equal to 100%.) It is interesting to note that the first two structures incorporate the resonance structure, and the third one was long ago postulated as an important contribution, for example by Bonino.

Let us now consider the hydrogen atoms. The designation for the hydrogen atoms are such that H_i is the hydrogen bonded to C_i. The energies for H_1-H_2, H_1-H_3 and H_1-H_4 are 0.4122 a.u., 0.2634 a.u. and 0.1107 a.u., respectively. As above done for the C-C bonds, each reported energy refers to the sum of the six possible H_1-H_2 type connections. The above data, for the two-center contributions, clearly indicates that each hydrogen atom has a repulsive interaction with each of the other hydrogen atoms.

Finally, let us consider the hydrogen-carbon interactions. For the traditional C-H bond (that is, C_1-H_1) there is a strong attraction (-4.5534 a.u.). The interaction is repulsive for all the remaining hydrogen-carbon pairs; namely, H_1-C_2, H_1-C_3 and H_1-C_4 yields 2.1024 a.u., 0.7296 a.u. and 0.4350 a.u., respectively.

We can now arrange the ten possible two-center BESs for benzene in order of importance. We obtain the results of Table 6. These results suggests the following analogy to the traditional structural formulae. There is a bond between C_i and H_i, and between C_i and C_{i+1}, in addition to the network of bonds C_i-C_{i+2} providing an interesting representation of aromatic structures.

If we consider the quantitative value of each BES independently from the sign, we conclude that the BESs of C_1--C_2 (bonding), H_1--C_1 (bonding), H_1--C_2 (antibonding), and C_1--C_4 type (antibonding) are the most important in benzene (but the remaining ones are not negligible). Since in our computations we have accounted for 71% of the experimental binding energy (the remaining 29% is partly the molecular extra correlation energy (22%), and partly the difference between our computed binding and the Hartree-Fock limit (7%)), we feel that the molecular binding mechanism here presented is very simple and in substantial agreement with the valence traditional concept, if we assume that, say for C_1, the C_1-C_3 and the C_1-C_5 semibonds correspond to the fourth valence electron (and bond) for the carbon atom. In the traditional chemical formulae, only the binding atom-atom connections are considered. Thus the above energy partition provides essentially the same information as available in the standard chemical formulae, but it adds to it the repulsive aspect for the two-centers and the energetic aspect for the one-centers. The inclusion of the three-and four-center terms will somewhat alter the above picture especially in the relative importance of the BES above reported. This type of analysis, including three and four-center bonds was lately abandoned in order to be closer to the traditional chemical formulae.

3.5 ORBITAL AND ELECTRON ENERGIES

Before proceeding with the bond energy analysis, it might be useful to introduce a few more definitions. Denoting the orbital energy associated with $\phi(i)$ by ε_i, it is easy to show (63) that the orbital energy can be partitioned into one, two, three and four-center components. Let us now define the following quantities:

$$\eta_i(A) = \varepsilon_i(A) + h_i(A)$$
$$\eta_i(A,B) = \varepsilon_i(A,B) + h_i(A,B)$$
$$\eta_i(A,B,C) = \varepsilon_i(A,B,C) + h_i(A,B,C) \qquad 3.3$$
$$\eta_i(A,B,C,D) = \varepsilon_i(A,B,C,D)$$

and let us indicate with η_i the sum of all such quantities belonging to $\phi(i)$, and name it the "electron energy".

Whereas the sum of the "orbital energies" is not equal to the total energy E, the sum of the "electron energies", η, is equal to the total energy. It requires, however, some attention for an open-shell structure. We note that the MOVS for an atom A is defined by the quantity:

$$E_A = \Sigma_i \eta_i(A) = \sum_i (\varepsilon_i(A) + h_i(A)) \qquad 3.4$$

Let us consider water as an example. A rather accurate Hartree-Fock computed energy yields the value of -76.06587 a.u. (68).

The ten electrons of H_2O, grouped into five molecular orbitals, have the orbitals energies given in Table 7; in this Table the orbital energies are decomposed into one, two and three-center contributions according to the BEA prescription (see Table 7).

Table 7. Orbital Energy Decomposition for H_2O (in a.u.)

Orbital Energy	\multicolumn{5}{c}{DECOMPOSITION*}				
	H(1)	O	H(1)--H(2)	H(1)--O	H(1)--H(2)--O
$\varepsilon_1 = -20.5604$	0.0	-20.8171	0.0	0.1502	-0.0437
$\varepsilon_2 = -1.3511$	-0.0033	-0.8431	-0.0211	-0.2392	-0.0017
$\varepsilon_3 = -0.7179$	-0.0374	-0.1426	0.0019	-0.3245	0.1465
$\varepsilon_4 = -0.5811$	-0.0110	-0.6091	-0.0428	0.0652	-0.0376
$\varepsilon_5 = -0.5082$	0.0010	-0.5837	-0.0005	0.0528	-0.0316

*The decomposition does not report H(2) and H(2)-O since equal to H(1) and H(1)-O, respectively.

The notation H(1)--O indicates a two-center contribution between H(1) and O; equivalently H(1)--H(2)--O, a three-center contribution. Table 7 clearly indicates that the first orbital is mainly due to the oxygen atom with non-negligible influence from the two-center contribution along the OH bond and a smaller three-center contribution. The zero contribution of H(1) (and of H(2)) indicates that there is no electronic charge on the hydrogens for this orbital. The above data are all in good agreement with our picture that the first orbital is essentially a $1s^2$ electron pair centered at the oxygen nucleus. It is noted that the orbital energy for the $1s^2$ electrons of the oxygen atom in the ground state configuration is -20.6686, -20.6932 and -20.7304 a.u. for the ^3P, ^1D and ^1S states, respectively (69). The value -20.8171 a.u. (see Table 7) reminds us that it takes energy to distort (even slightly) an orbital; the total orbital energy in H_2O of -20.5604 a.u. is nearer to the value for the ^3P state, but this value should not be interpreted as "nearly the same energy as in the atomic ground state because of zero molecular field" since the value -20.5604 a.u. is obtained by summing the molecule field contribution (it represents the balance of opposing effects).

As known, the second orbital is mainly a $2s^2$ on the oxygen, whereas the third and fourth are mainly the bonding orbitals between O and H, and the fifth orbital is mainly a lone pair electron of $2p_z$ type, perpendicular to the molecular plane. The orbital energy for the $2s^2$

electrons in the oxygen atom is -1.2443, -1.2565 and -1.2751 a.u. for ^3P, ^1D, and ^1S states, respectively. The orbital energy of the second orbital in H_2O is -1.3511 a.u., but its decomposition reveals that the near equality to the value of -1.25 a.u. of the separated atom is the result of a rather drastic reorganization (with respect to the separated atom), whereby the H--O term is significantly important. Table 7 warns us that to equate the orbital energy of separated atoms with the orbital energies of a molecule can be misleading, if the near equality in the energy is interpreted as no molecular field effects. The two bonding orbitals (third and fourth) are close in energy but very different in their compositions. The dominant terms of the third orbital are the two two-center H--O terms, whereas the dominant term in the fourth orbital is associated with the one-center energy (on the oxygen atom).

From Table 7 the water molecule appears to be a deformed O^{2-} ion with two protons. The five orbitals are $\phi_1 = 1s^2$ (0), $\phi_2 = 2s^2$ (0), $\phi_3 = 2p_x^2$ (0), $\phi_4 = 2p_y^2$ (0) and $\phi_5 = 2p_z^2$ (0).

Table 8. Electron Energy Decomposition for H_2O (in a.u.)

Electron Energy	H(1)	O	DECOMPOSITION H(1)--H(2)	H(1)--O	H(1)--H(2)--O
$\eta_1 = -53.6088$	0.0	-52.7277	0.0	-0.4086	-0.0439
$\eta_2 = -9.2098$	-0.0111	-0.8925	-0.0501	-0.3754	-0.3241
$\eta_3 = -7.4404$	-0.0959	-3.1344	-0.0077	-2.2463	0.3859
$\eta_4 = -7.4692$	-0.2917	-6.0643	-0.1016	-0.5376	-0.1798
$\eta_5 = -7.5441$	+0.0018	-6.4608	-0.0012	-0.6234	-0.0467

This crude picture does not account for the nonlinear structure of water namely, it would be more consistent with a linear H_2O. In Table 8, we report the breakdown of the η's in terms of one, two and three-center components.

The first point we notice in Table 8 is that the order of the three orbitals ϕ_3, ϕ_4, ϕ_5 is inverted; ϕ_5 is the most energetic, ϕ_4 the second and ϕ_3 the least energetic. If we ask for the energy of an electron in the molecule, then Table 8 is at least as useful as Table 7. We feel that correlation diagrams between electron energies should complement and at times substitute the standard correlation diagrams between orbital energies, especially in connection with interpretation of chemical reactions. Indeed, by using the orbital energies rather than the electron energies, the role of the Kinetic energy and of the nuclear-electron attraction is less clear and only about halfway considered. The reversal of the η_5 relative to the ε_3 is not surprising: our rough model which equates H_2O with O^{-2} plus two protons, has to add to the perturbed ^3P state of oxygen the two electrons of the two hydrogens in order to construct an additional 2p orbital. Since ϕ_3 is the "most distorted 2p orbital" of the three pairs of orbitals, and ϕ_5 is the least distorted, then, because "distortions" cost energy, ϕ_5 is the most stable, ϕ_3 the least stable, and ϕ_4 is of intermediate stability relative to the two. The second point to notice is that the last three orbitals are nearly equivalent in binding energy (about 1% difference between ϕ_3, ϕ_4, and ϕ_5), consistent with the "O^{2-} plus two protons" picture. In Table 8, η_3 is mainly formed by the O--H two-center terms (-0.6490 a.u. compared to $\varepsilon_3 = -0.7179$ a.u.); in Table 7, η_3 is mainly the sum of both the one-center term at the O site, and the two two-center terms of O--H type. The main repulsive term in both Tables 7 and 8 is the three-center term H--O--H in ϕ_3. Tentatively, we shall postulate that whereas the two-center terms are of predominant importance in defining bond lengths, three-center terms are predominantly important for bond angles.

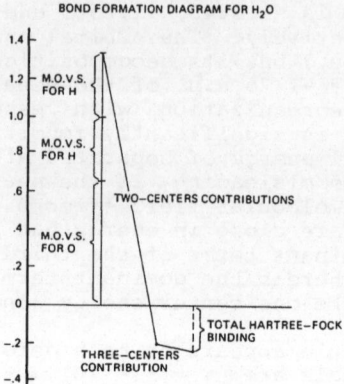

Figure 7. Bond-formation diagram for H_2O. The one-center energies (MOVS) require about 1.3 a.u. of energy from the atoms in the ground state (zero of figure); the two-center energies more than compensate this loss of energy, and the three-center energies, refine somewhat the overall situation.

The bond formation diagram for the H_2O molecule in the Hartree-Fock approximation is given in Figure 7. At zero energy we take the sum of the Hartree-Fock energies of oxygen in the ^3P state (-74.8093 a.u.) and hydrogen in the ^2S state (-0.5000 a.u.). The one-center energies (for H_2O) summed over the five orbitals are -74.1525 and -0.1342 a.u. for the oxygen and for each hydrogen atom, respectively. Thus, the one-center sum is 1.3884 a.u. above the above-defined zero. The total two-center energies are 0.1880 a.u. for H--H and -0.8896 a.u. for each of the O--H, or a total of -1.5912 a.u. The two-center contribution to the total energy, therefore, compensates for the losses in energy of the molecular orbital valency state and brings about a binding of 1.5912 -1.3884 = 0.2028 a.u. The three-center energy increases the binding by 0.0434 a.u., yielding a total Hartree-Fock binding energy of 0.2562 a.u.

3.6 MOVS AND HYBRIDIZATION

Let us consider a different set of examples, namely the molecules CH_4, CH_3F, CH_2F_2, CHF_3 and CF_4 (63).

We shall now analyze the gross behavior of the MOVS in the compounds CH_4, CH_3F, CH_2F_2, CHF_3, and CF_4. It is noted that the wave functions considered here do not represent the Hartree-Fock limit; on the other hand, they are of sufficient accuracy to indicate trends correctly.

In Table 9 we report the MOVS for the carbon, hydrogen and fluorine atoms. From this table we learn once more that the MOVS is strongly dependent on the environment. It ought to be so. Table 10 reports the gross atomic charges for the above compounds. By comparing the data of Table 9 with those of Table 10, we learn that the MOVS energy is nearly a regular function of charge-transfer, and the more an atom acts as an acceptor, the less unstable the MOVS becomes relative to the ground state energy of the separated atoms. As previously pointed out,

Table 9. MO valency state for the carbon, fluorine and hydrogen atoms as from bond energy analysis (in a.u.) (*)

Molecule	C	F	H
CH_4	-36.6266	---	-0.1994
CH_3F	-36.4193	-99.2982	-0.2053
CH_2F_2	-36.1097	-99.2627	-0.2124
CHF_3	-35.7293	-99.2411	-0.2048
CF_4	-35.3239	-99.2296	---

(*) Valency state standard for the tetrahedral carbon: -37.4585 a.u.

Table 10. Gross Charges and Hybridization

Orbital	CH_4	CH_3F	CH_2F_2	CHF_3	CF_4
1s(H)	0.826	0.780	0.832	0.823	---
1s(C)	2.000	2.000	2.000	2.000	2.000
2s(C)	1.452	1.346	1.209	1.057	0.910
2p(C)	3.261	2.782	2.435	2.222	2.083
1s(F)	---	2.000	2.000	2.000	2.000
2s(F)	---	1.933	1.942	1.938	1.938
2p(F)	---	5.355	5.404	5.361	5.314

the MOVS can be more stable than the separated atom in the ground state for the case of strongly electronegative atoms like fluorine when in an appropriate environment. We recall that the Hartree-Fock energy for the fluorine atom in the ^2P state is -99.4093 a.u., and the equivalent quantity for the carbon atom is -37.6886 a.u. The energy difference between MOVS energy and the ground state for the separated atom provides the basis parameter to define an electronegativity scale atoms in molecules.

3.7 BOND ENERGY ANALYSIS: A "NEW" FORMALISM

In the two previous sections, we have decomposed the total energy of a system using the bond energy analysis as presented in the early stages of its development.

In this section we shall derive a simpler energy partitioning that is intended to provide a parallelism as near as possible to some intuitive idea about chemical bonds. Preliminary accounts have been previously reported (70). The standard hamiltonian for the m,n-system yields the by now familiar one, two, three, and four-center contributions to the total energy. In more detail, the kinetic part yields one and two-center terms; the nuclear-electron attraction part yields one, two and three-center terms; the electron-electron part yields one, two, three, and four-center terms; the nuclear-nuclear repulsion yields two-center terms.

To map these terms into two-center terms, the necessary condition for obtaining an analog to the classical chemical formulae, we shall make use of a number of transformations described below. To simplify the notation, we shall write a basis set χ_A centered on the atom A simply as A; with this notation we describe the following transformation rules:

1. Kinetic matrix elements: the $\langle A|\nabla|A\rangle$ matrix elements are assigned to an energy $E(A)$ and the $\langle A|\nabla|B\rangle$ matrix elements are assigned to an energy $E(AB)$;

2. Nuclear-electron matrix elements: the $\langle A|Z_A/R|A\rangle$ elements and the $\langle A|Z_B/R|A\rangle$ elements are assigned to the energy $E(A)$; the $\langle A|Z_A/R|B\rangle$ elements, the $\langle A|Z_B/R|B\rangle$ elements are assigned to the energy $E(AB)$;

3. Electron-electron matrix elements: the terms $\langle AA|1/r|AA\rangle$ are assigned to the energy $E(A)$: the terms $\langle AA|1/r|AB\rangle$, $\langle AA|1/r|BB\rangle$, $\langle AB|1/r|AB\rangle$ are assigned to the energy $E(AB)$: the terms $\langle AB|1/r|CD\rangle$ are assigned to $E(AC)$, $E(AD)$, $E(BC)$ and $E(BD)$ with the transformation relation (used also for three-centers):

$$\langle AB|1/r|CD\rangle = W(\langle AA|1/r|CC\rangle + \langle AA|1/r|DD\rangle + \langle BB|1/r|CC\rangle + \langle BB|1/r|DD\rangle)$$

The weight, W, is given by the relation:

$$1/W = (\langle AA|1/r|CC\rangle + \langle AA|1/r|DD\rangle + \langle BB|1/r|CC\rangle + \langle BB|1/r|DD\rangle)/\langle AB|1/r|CD\rangle$$

4. Nuclear-nuclear repulsion: the $(Z_A Z_B/R)$ energy is obviously assigned to the $E(AB)$ energy term.

Therefore, with the above transformation rules the total energy E corresponding to a wave function describing m nuclei and n electrons is decomposed into pair-wise energies;

$$E = \sum_{AB} E(AB) = \sum_A E(A) + \sum_A \sum_{A \neq B} E(AB) \qquad 3.5$$

where the index A (or B) runs over all the atoms of the molecule.

Let us now consider a set of four atoms designated as 1, 2, 3, and 4 and let us assume that two of them (1 and 2) form a molecule and the second pair (3 and 4) form a second molecule. We consider three conformations for the four atoms: 1) all four atoms are infinitely far from each other, 2) atoms 1 and 2 (as well as 3 and 4 are at a distance corresponding to a finite nonzero interaction (for example, the equilibrium distance) but the two molecules are infinitely separated, and 3) as in the previous case, but with the two molecules interacting with each other.

Using (3.5) for the three cases we obtain three matrices designated as A, B', and C', described below:

Matrix A

	1	2	3	4
1	a(11)	0	0	0
2	0	a(22)	0	0
3	0	0	a(33)	0
4	0	0	0	a(44)

Matrix B'

	1	2	3	4
1	b'(11)	b'(12)	0	0
2	b'(21)	b'(22)	0	0
3	0	0	b'(33)	b'(34)
4	0	0	b'(43)	b'(44)

Matrix C'

	1	2	3	4
1	c'(11)	c'(12)	c'(13)	c'(14)
2	c'(21)	c'(22)	c'(23)	c'(24)
3	c'(31)	c'(32)	c'(33)	c'(34)
4	c'(41)	c'(42)	c'(43)	c'(44)

We note that the matrix A is defined as that matrix having all elements except the diagonal ones equal to zero, the B' matrix is the one having a symmetric number of terms (b'(ij) and b'(ji)) equal to zero, the matrix C' is the one having no pair of symmetric terms equal to zero; the matrices A, B' and C' are square and symmetric matrices. By subtracting each element a(ij) of A, from the corresponding element b'(ij) of B', one obtains the total binding energy for the two separated molecules; in this operation we subtract from the molecular energy terms (matrix B') the energy of the separated atoms (matrix A).

By subtracting each element a(ij) of A from the corresponding element c'(ij) of C' one obtains the total binding energy for the complex.

By subtracting each element b'(ij) of B' from the corresponding element c'(ij) of C' one obtains the interaction energy of the two molecules relative to the energy of the two molecules at infinite separation. Clearly, if the four atoms (or n atoms) do not form two distinct molecules (as in our example) but a single molecule, then only the matrix A and the matric C' are of interest.

The matrices A, B' and C', above defined have an immediate physical interpretation; as pointed out, however, such matrices suffer from the fact that the diagonal terms are nonzero and therefore cannot be used directly to provide a quantitative model for classical chemical structural formulae, where there is no self-energy for the atoms. In the following we shall distribute the atomic energy (one-center, that is, the diagonal terms) on the nondiagonal terms.

Let us now define a matrix B of elements b(ij), a matrix C of elements c(ij) and a matrix IP, of elements designated either as p(ij) or I(ij); the construction rules are as follows:

$$b(ij) = 1/2[b''(ij) + b''(ji)] \quad ; \quad c(ij) = 1/2[c''(ij) + c''(ji)] \quad \quad 3.6$$

where
$$b''(ij) = [1-\delta(ij)] \{ b'(ij) + [b'(jj) - a(jj)] * \\ *\{|b'(ij)|/\Sigma_j|b'(ij) - [1-\delta(ij)]|\}\} \quad \quad 3.7$$

and
$$c''(ij) = [1-\delta(ij)] \{c'(ij) + [c'(jj) - a(jj)] * \\ *\{|c'(ij)|/\Sigma_j|c'(ij) - [1-\delta(ij)]|\}\} \quad \quad 3.8$$

$$p(ij) = c(ij) - b(ij) \quad ; \quad I(ij) = c(ij) \quad \quad 3.9$$

In our example the pair indices i and j are (1,2) and (3,4) for p(ij), and (3,1), (3,2), (4,1) and (4,2) for I(ij). Therefore we have:

Matrix B

	1	2	3	4
1	0	b(12)	0	0
2	b(21)	0	0	0
3	0	0	0	b(34)
4	0	0	b(43)	0

Matrix C

	1	2	3	4
1	0	c(12)	c(13)	c(14)
2	c(21)	0	c(23)	c(24)
3	c(31)	c(32)	0	c(34)
4	c(41)	c(42)	c(43)	0

Matrix IP

	1	2	3	4
1	0	p(12)	I(13)	I(14)
2	p(21)	0	I(23)	I(24)
3	I(31)	I(32)	0	p(34)
4	I(41)	I(42)	p(43)	0

The matrix B represents the energy of the BEA-bond of the first molecule (namely 2b(12)) and the energy of the BEA bond of the second molecule namely 2b(34); the matrix C represents all the pair-wise interactions for the two molecules when interacting. Some of the matrix elements are repulsive (positive values), some are attractive (negative values). Among the attractive ones we select the most important (see below) and call each of them a BEA bond; the remaining correspond to the BEA nonbonded interactions. The notation C stands for "connection" matrix. The matrix IP represents the interaction of the two molecules expressed as the variation (by polarization and induction and other terms) of the energy of the BEA bonds of one molecule due to the field of the second (p(12) and p(21)), the variation of the energy of the BEA bonds of the second molecule due to the field of the first one (p(34) and p(43)) and the nonbonded interactions denoted as I(ij). Clearly, if the system is a single molecule, then the matrix IP does not exist.

If we indicate with $\mathscr{E}(M)$ and $\mathscr{E}(N)$ the sums of the energies of the atoms constituting the molecules M and N, respectively, with E(M) and E(N) the total energies of the molecules M and N and with E(M,N) the total energy of the complex M-N, then the following equalities hold:

$$\sum_{i,j=1}^{2} a(ij) = \mathscr{E}(M) \quad \text{and} \quad \sum_{i,j=3}^{4} a(ij) = \mathscr{E}(N) \qquad 3.10$$

$$\sum_{i,j=1}^{2} b'(ij) = E(M) \quad \text{and} \quad \sum_{i,j=3}^{4} b'(ij) = E(N) \qquad 3.11$$

$$\sum_{ij} c'(ij) = E(MN). \qquad 3.12$$

The binding energy of M expressed as the sum of all bonded and non-bonded interaction is:

$$\sum_{i,j=1}^{2} b(ij) = E(M) - \mathscr{E}(M) \qquad 3.13$$

The binding energy of N expressed as the sum of all bonded and non-bonded interactions is:

$$\sum_{i,j=3}^{4} b(ij) = E(N) - \mathscr{E}(N) \qquad 3.14$$

The binding energy of the complex M-N expressed as the sum of all the bonded and nonbonded interactions is:

$$\sum_{ij} c(ij) = E(M,N) - E(M) - E(N) \qquad 3.15$$

The polarization and induction energies of M due to N is $\sum_{i,j=1}^{2} p(ij)$ and the polarization energy of N due to M is $\sum_{i,j=3}^{4} p(ij)$; finally, the sum of nonbonded interactions between M and N in the complex is $\sum_{i=1}^{2} \sum_{j=3}^{4} I(ij)$.

The extension to more than two molecules is trivial. It is stressed that these definitions hold not only for any function in the SCF-LCAO-MO approximation, but also for CI wave functions or any wave function expressed as a linear combination of Slater determinants.

3.8 CHEMICAL FORMULAE FROM THE BONDED ATOM PAIRS ANALYSIS

From the matrix C we can obtain by inspection the traditional chemical formula, namely we can associate the traditional presence of a bond to selected pairs of atoms (this is not the same as saying that the traditional bond is equated to a pair interaction). The rule is very simple: no term with repulsive interaction corresponds to a bond and an attraction larger than a given threshold indicates the existence of a traditional bond and is referred to as a BEA-bond. Alternatively, we order the matrix element from the most attractive to the most repulsive, and we select, starting from the first term of the ordered set, as many connections as needed to satisfy the valence requirements. In practice both criteria are used and a minimum of good common sense is required.

If one wishes to obtain a more stringent parallelism between the classical representation of a chemical bond, as standardly used for structural formulae, and BEA bonds, it is possible to eliminate all the nonbonded interactions of intramolecular type (not present in the classical representation). To perform this task one distributes each of the nonbonded intramolecular interactions onto that subset of the BEA-bonds that are in the "physical space" of the nonbonded interactions. For example, a nonbonded intramolecular interaction between atoms A and B and with intermolecular separation R(AB) can be partitioned with a simple weighting algorithm on the subset of BEA-bonds contained in the space of a sphere of radius proportional to R(AB). In this way the intramolecular nonbonded interaction is distributed on the bonds, and, with a minimum of care, one can ensure that the nonbonded interaction of a functional group is distributed on the bonds of the functional group. This new representation is called **BAP representation**, "bonded atom pairs".

Let us consider more formally the algorithm needed to pass from the BEA representation to the BAP representation. On the base of the threshold energy criterium, we know that for our m atoms systems, only NB bonds are present among the $m(m-1)/2$ possible connections. Let us use the indices t and u for atom pairs connected by BEA-bonds, and the k and l indices for atom pairs connected by nonbonded interactions. The BAP can be written in the usual half-matrix form, with elements $d(t,u)$ with $t=1,\ldots,NB$ and $u \leq t$. Then for BAP-bonds we have

$$d(t,u) = c(t,u) + \Sigma\Sigma w(k,l,t,u) c(k,l) \qquad 3.16$$

and for the nonbonded interactions

$$d(k,l) = c(k,l) - c(k,l) = 0 \qquad 3.17$$

where with the weight $w(k,l,t,u)$ we add to the bond interaction $c(t,u)$ a fraction of each of the nonbonded interactions $c(k,l)$. The matrix elements corresponding to intermolecular interaction are left unchanged; however, it is simple to redistribute them on the bonds, if desired. (In the example below analyzed, the intermolecular interactions in the D matrix are left as in the C matrix.) The explicit form for the weight factor is obtained by considering a sphere with the center at the midpoint of the nonbonded interaction $c(k,l)$, namely at the midpoint of the distance between the k-th and the l-th atoms. Such sphere will contain a number of BEA-bonds, and an equal number of midpoints for each of the BEA-bonds. Thus, we can define the distance from the sphere cluster for each BEA-bond in the sphere. The weight factors are proportional to such distance, namely more of $c(k,l)$ goes to the nearest BEA-bond.

In tables 11 to 15 we report the matrices B', C', B, C and D obtained in the BEA and in the BAP formalism for two interacting molecules CH_4 and H_2O. The indices representing our m=8 atoms are 1 for the carbon atom, 2, 3, 4 and 5 for the hydrogen atoms of CH_4, 6 for the oxygen atom and 7 and 8 for the hydrogen atoms of H_2O. The hydrogen 7 and 8 are symmetrically located relative to the C-H...O axis and the OH bonds point away from CH_4; the hydrogen 2 is the one on the C-H...O axis.

Table 11. Matric $\underline{B}$' for the $CH_4 + H_2O$ Study (in a.u.)

	1	2	3	4	5	6	7	8
1	0.997							
2	-0.841	0.291						
3	-0.841	0.068	0.291					
4	-0.841	0.068	0.068	0.291				
5	-0.841	0.068	0.068	0.068	0.291			
6	0.0	0.0	0.0	0.0	0.0	0.433		
7	0.0	0.0	0.0	0.0	0.0	-0.692	0.294	
8	0.0	0.0	0.0	0.0	0.0	-0.692	0.191	0.294

Table 12. Matrix $\underline{C}$' for the $CH_4 + H_2O$ study (in a.u.)

	1	2	3	4	5	6	7	8
1	0.983							
2	-0.844	0.309						
3	-0.821	0.063	0.285					
4	-0.826	0.063	0.639	0.285				
5	-0.826	0.063	0.639	0.864	0.285			
6	0.117	-0.068	-0.021	-0.021	-0.021	0.445		
7	-0.042	0.018	0.008	0.009	0.009	-0.700	0.280	
8	-0.042	0.018	0.009	0.009	0.009	-0.700	0.190	0.280

Table 13. Matrix $\underline{B}$ for the $CH_4 - H_2O$ study (in a.u.)

	1	2	3	4	5	6	7	8
1	0.0							
2	-0.253	0.0						
3	-0.253	0.104	0.0					
4	-0.253	0.104	0.104	0.0				
5	-0.253	0.104	0.104	0.104	0.0			
6	0.0	0.0	0.0	0.0	0.0	0.0		
7	0.0	0.0	0.0	0.0	0.0	-0.215	0.0	
8	0.0	0.0	0.0	0.0	0.0	-0.215	0.312	0.0

Table 14. Matrix C for the $CH_4 + H_2O$ Study (*)

	1	2	3	4	5	6	7	8
1	0.0							
2	-0.277	0.0						
3	-0.275	0.094	0.0					
4	-0.278	0.094	0.096	0.0				
5	-0.278	0.094	0.096	0.103	0.0			
6	0.203	-0.028	-0.009	-0.009	-0.009	0.0		
7	-0.012	0.028	0.013	0.013	0.013	-0.271	0.0	
8	-0.012	0.028	0.013	0.013	0.013	-0.271	0.303	0.0

(*) Matrix elements in a.u.

Table 15. BAP interactions for CH_4+H_2O - the D matrix (in a.u.)

	1	2	3	4	5	6	7	8
1	0.0							
2	-0.136	0.0						
3	-0.132	0.0	0.0					
4	-0.125	0.0	0.0	0.0				
5	-0.125	0.0	0.0	0.0	0.0			
6	0.203	-0.028	-0.009	-0.009	-0.009	0.0		
7	-0.012	0.028	0.013	0.013	0.013	-0.120	0.0	
8	-0.012	0.028	0.013	0.013	0.013	-0.120	0.0	0.0

The carbon-oxygen separation is 6.0 a.u. The matrix A (atoms at infinite separation) consist of the ordered diagonal elements -37.609 a.u., -0.497 a.u., -0497 a.u., 0-.497 a.u., -0.497 a.u., -74.621 a.u., -0.497 a.u. and -0.497 a.u. All the remaining nondiagonal elements being zero; these are the total energies for the carbon atom in the ground state, for the four hydrogen atoms, for the oxygen atom and for the two remaining hydrogen atoms. The basis set is our standard minimal basis (7s,3p) and the computations are limited to the S.C.F. level. In the A matrix the atoms are at infinite separation and thus noninteracting one with another. The matrix B' (Table 11) is given for CH_4 and for H_2O at the equilibrium distance in a configuration where CH_4 is infinitely distant (noninteracting) with H_2O.

From the matrix B' we learn at once that the molecular orbital valency states energies (the diagonal elements) are large and repulsive for the C, the H and the O atoms in CH_4 and H_2O. We also learn that it requires slightly more energy to deform the hydrogen from the $2s$ state to obtain the best distribution for CH_4 than for H_2O. Simply by inspection of the B' matrix we can anticipate that there are bonds between the carbon and the hydrogen atoms in the CH_4 (elements (2,1), (3,1), (4,1) and 5,1)) and that the H-H interactions are repulsive. Equivalently, simply by inspection of the B' matrix, we can anticipate that there are bonds between the oxygen atom and the hydrogen atoms in H_2O (elements (6,7) and (6,8)) whereas the hydrogen hydrogen interaction is repulsive (non-bonded) and more so than in the CH_4 molecule. In Table 12 we provide the matrix C'; here we can note that the CH_4 interaction with H_2O, at the chosen configuration, brings about small variations in the bonding and in the nonbonding intramolecular interactions and in the molecular orbital valence state energies, as one would expect. The intermolecular interaction energies are attractive between the oxygen atom of H_2O and the hydrogen atoms of CH_4 (elements (6,2), (6,3), (6,4) and (6,5)) and between the carbon atom in CH_4 and the hydrogen atoms in H_2O (elements (7,1) and (7,2)); all the hydrogen-hydrogen intermolecular interactions are repulsive.

In all the above matrices the atomic self-energy is nonzero; to be more in tune with the tradition of the bond, we pass now to the B matrix and the C matrix representing the CH_4+H_2O system with the two molecules either at infinite separation or at the finite separation before described. The corresponding matrices are given in Tables 13 and 14. The main aspect of these matrices is that the bonded and nonbonded interactions are now much smaller than in the B' and C' matrices, since the large repulsion of the molecular orbital valency state energy has been distributed into the atom-atom pair attractions. The numerical values for the bond are however still larger than those we generally associate with empirical bond strength, and the reason is that the nonbonded repulsion is large and not repartitioned into the interactions of bonding type.

As the last step we provide the BAP energies; namely, the D matrix, given in Table 15. Here the only nonzero elements corresponds to bonds and to intermolecular interactions; thus we have obtained the desired correspondence with traditional chemical formulae.

3.9 DEFINITION OF ATOMS AND MOLECULES

Having discussed an energy partitioning, we recall again that the partitioning of the total density into charge distribution around atoms is a very familiar concept since the original work of Mulliken (1), or even since Lewis' time.

In the following, however, we shall introduce a definition of atoms and molecules primarily based on the energetic aspect of a bond (71).

Let us start with the m,n-system and compute both E and its density distribution $\rho(x,y,z)$. Let us partition the energy into $m(m-1)/2$ parts, designated as $c'(a,b)$ (namely the elements of the C' matrix) such that:

$$E = 2\sum_{a=1}^{m} \sum_{b \neq a}^{m} c'_{ab} + \sum_{a=1}^{m} c'_{aa} \quad \text{and} \quad \rho = \sum_{a=1}^{m} \rho(a) \qquad 3.18$$

Let us now define m objects (that is as many objects as the nuclei of the m,n-system) and assign to the i-th object (i=1,...,m), a nuclear charge Z_i, the nuclear position, x_i, y_i, z_i, the density distribution corresponding to the gross atomic charge, ρ_i, and the subset (of length m-1) of the partitioned energies defined by the terms $c'(i,j)$, (j=1,......,m). If we designate the m objects as $A_1,...A_m$, we can write:

$$A_1 = Z_1, X_1, Y_1, Z_1, \rho_1, C'_{11}, C'_{12},....,C'_{1m} \qquad 3.19$$
$$A_2 = Z_2, X_2, Y_2, Z_2, \rho_2, C'_{21}, C'_{22},....,C'_{2m}$$
$$\cdots\cdots\cdots\cdots\cdots\cdots\cdots\cdots\cdots\cdots\cdots\cdots\cdots\cdots$$
$$A_m = Z_m, X_m, Y_m, Z_m, \rho_m, C'_{m1}, C'_{m2},...., C'_{mm}$$

This collection represents a first simplification and organization of the entire system of particles since from a system of m + n <u>particles</u> we pass to a system of m <u>objects</u>. The quantities C'_{ij} are real numbers: positive, zero or negative. We define that a bond between two objects A_i and A_j exists if the value of C'_{ij} is smaller (more negative) than a pre-assigned threshold; we define C'_{ij} as the "strength index" of that bond. We define as "nonbonded interactions" all the C'_{ij}s that are not "bonds". We define a nonbonded interaction as attractive or repulsive according to its sign (negative for attractive, positive for repulsive). Within the set of m objects, we select K subsets that must fulfill the following condition: all the elements of a subset must have at least one bond with another element of the subset. If one object has all C'_{ij} values equal to zero, then it is called a "noninteracting" member in the set of objects (above, the term "zero" is taken in the physical and not in the mathematical sense).

We define a molecule as the subset composed of two or more elements; we define an atom as a subset with only one element, namely an element of noninteracting type.

We define as a positive or negative ion the subset where the sum of the differences $Z_i - \rho_i$ for all its members is either positive or negative.

We define as intramolecular interaction the C'_{ij} within the elements of the same set; we define as intermolecular interactions the C'_{ij} between elements of two or more subsets.

Thus an object A_i is bonded or nonbonded to an object A_j depending on the C'_{ij} value; in addition, the object A_i has a different stability in the system relative to its being at infinite separation from the remaining objects; the difference in stability is given by $\Delta E = C_{ii} - E_{jj}$, where E_{jj} is the ground state energy (namely, the energy of the object when it is a noninteracting element of the set).

We note that a chemical structural formula as obtained from chemical tradition is in satisfactory agreement with the above definitions derived directly from time-independent quantum mechanics; the above definition constitutes only a partial definition of a chemical bond since we have neglected the electronic density aspect and concentrated mainly on topological and energetic factors. For additional observations on this section, we refer to some earlier work (60).

We note in addition that our definition of bond existence relies on a threshold value; we comment that part of the history of chemistry is related to changes in the commonly accepted threshold value. The evolution from the strong and classical bonds, to the hydrogen bond, to bonds between rare gases and halogens, and more recently to Van der Waals bonds can be translated into variation of accepted threshold value.

Let us now extend the above definition of a molecule, by making more use of the second characterization of a bond; namely, of its density distribution. In the previous definition we have essentially made use of the C' matrix (BEA representation); in the following, we shall use mainly the D matrix (BAP representation). We assume to have at our disposal the D matrix and the Mulliken's gross population on each atom; namely, quantities derived directly from the system of m nuclei and n electrons and from approximations to the Schroedinger equation (as those based on the Born-Oppenheimer approximation); explicit reference to the correlation correction is a nonessential point in what follows, since both the D matrix and the gross populations can be obtained either in the SCF or in any type of MC-SCF approximations. We redefine the set of m objects by associating to each object, for example, the i-th one, a nuclear charge Z_i, a space identification x_i, y_i and z_i, the $d(i,j)$ matrix elements of the D matrix (j runs from 1 to m) and electronic density characterizations, as explained below.

The total electronic population ρ of the m atoms can be partitioned into atomic gross populations on each atom of the familiar type $1s^r$, $2s^s$, $2p_x^t$, $2p_y^u$, $2p_z^v$ ...; in turn, these populations can be partitioned into inner shell electrons ρ^0 and valency electrons ρ', the distinction being rather immediate upon inspection, case by case. Let us indicate as $\rho(i)$ the valency electron gross population for the i-th atom and let us assume that it is expressed into components ordered by molecular symmetry rather than by atomic symmetry; we use the index μ for the symmetry species and the index λ to indicate terms of given symmetry. For example, if in the above familiar atomic distribution, the molecular symmetry is σ and π, then we can use the indices $\mu=1$, $\lambda=1$ for 2s, $\mu=1$, $\lambda=2$ for $2p_x$, $\mu=1$, $\lambda=3$ for $2p_y$ and $\mu=2$, $\lambda=1$ for $2p_z$ (the $2p_z$ is taken as the $2p\pi$ term). Thus we will use the general (neglecting the inner shell 1s electron) notation $\rho(i_\lambda^\mu)$ of $2s^s = \rho(\begin{smallmatrix}1\\1\end{smallmatrix})$, $2p_x^t = \rho(\begin{smallmatrix}1\\2\end{smallmatrix})$, $2p_y^u = \rho(\begin{smallmatrix}1\\3\end{smallmatrix})$ and $2p_z^v = (\begin{smallmatrix}2\\1\end{smallmatrix})$; to distinguish the atom, we add its index (either i or j, where $i,j=1,...,m$). The total electronic density for the valence electrons in the molecule is given by $\rho' = \sum_i \rho(i_\lambda^\mu)$.

From the $d(i,j)$ matrix elements, we know how many bonds are departing from the i-th nucleus. Let us designate the corresponding number by $b(i)$. We can partition the $\rho(i_\lambda^\mu)$ populations on the $b(i)$ bonds, and attribute to each bond a fraction of the population as $w(i_\lambda^\mu, k) \rho(i_\lambda^\mu)$

such that

$$\rho(i_\lambda^\mu) = \sum_{k=1}^{b(i)} w(i_\lambda^\mu, k) \rho(i_\lambda^\mu) \qquad 3.20$$

The total population along the i-j bond will be given by the quantity

$$\rho(i,j) = \sum_{\mu\lambda}[w(i_\lambda^\mu,j)\rho(i_\lambda^\mu) + w(j_\lambda^\mu,i)\rho(j_\lambda^\mu)] \qquad 3.21$$

There are a number of ways to determine the weight factor w. For example, one can perform a numerical integration on some region of space assumed to represent a bond, for example, on the base of topological properties of the electronic density. Indeed one could use techniques of the type proposed by R. Bader (57); likely, however, we might prefer some method computationally less expensive and capable of distinguishing between H_2 and He_2, namely between bonding and antibonding situations. In this respect, the intersecting sphere approximation introduced by S. Antoci (19) might be most valuable. For a preliminary analysis we can use some very simple partitioning technique, that does not even require the use of computers as shown below.

For example, let us consider the pyridine molecule, using the wave function results reported (72) several years ago (1967). The electron's population is partitioned by projection of the atomic densities either along the x or the y axis (the molecule is in the x-y plane, with the y-axis bysecting the nitrogen atom and the opposing carbon atom designated as C(5)). First we combine the 2s and the $2p_y$ populations on the nitrogen atom, yielding a population 1.50896 + 1.66076 = 3.16972 electrons; performing, equivalently, for C(5) we obtain a population of 2.21666 electrons. This mixing is allowed because the 2s and the $2p_y$ populations belong to the same symmetry. We then equally subdivide the 2s populations for each one of the remaining atoms into the $2p_x$ and $2p_y$ populations yielding a mixed population (2s+2p$_x$) equal to 1.56532 electrons for C(1) or C(2), the two carbon atoms neighboring the nitrogen atom, and 1.64671 electrons for C(3) or C(4), the two carbon atoms neighboring C(5), and a mixed population (2s+2p$_y$) equal to 1.53667 electrons for C(1) or C(2) and 1.57693 for C(3) or C(4). At this point we project the above populations <u>onto the bond directions determined by the D matrix</u>, keeping in mind, however, the existence of the nitrogen lone pair (along the y axis). For the nitrogen atom, the (2s+2p$_y$) populations above obtained are subdivided into the three bonds (the lone pair is counted as a bond). Because of the molecular symmetry, half population is assigned to bonds in the y directions, and one quarter of the remaining population is attributed to each of the bonds not along the y direction and departing from the atom in consideration. The (2s+2p$_x$) populations are subdivided equally into the two bonds that are not along the y direction. The $2p_x = 2p_\pi$ electrons are clearly unmixed. In this way the lone pair population results in 1.58 electrons, the N-C(1) populations (on the σ bond) is 2.48 electrons, 1.32 coming from the N atom and 1.17 coming C(1). The σ-type population on the C(1)-H(1) is 1.94 electrons, 0.78 from H(1) and 1.17 from C(1). The σ-type population on the C(1)-C(3) is 1.56 electrons, 0.77 from C(1) and 0.79 from C(3); the C(3)-H(3) bond has 2.00 electrons, 1.22 from C(3) and 0.78 from H(3); the C(3)-C(5) bond has 2.28 electrons, 1.22 from C(3) and 1.06 from C(5); finally, the C(5)-H(5) has 1.89 electrons, 1.11 from C(5) and 0.78 from H(5). The π electron's distribution consists of essentially one electron per π bond (more precisely, a little less for the C(5)-C(3) or C(5)-C(4) bonds).

The weights here selected are particularly simple in order to perform an elementary handcomputation, and, as a consequence, the analysis of the electronic population distribution, bond by bond is approximated.

A more accurate analysis, however, will bring out the same features with some electronic charges, possibly, moving from the N-C(1) and C(3)-C(5) bonds to the C(1)-C(5) and C(2)-C(4) bonds.

In conclusion, it appears that today we can state a few of the basic characterizations of a bond in addition to those reported in the early results concerning pyridine (72) and of those obtained from the C or the D matrices. At the time of the pyridine study (72) <u>all electrons</u> ab initio computations where still strongly disdained by a number of quantum chemists. Thus, it was important to show that the two-way σ,π charge transfer is a basic aspect of the molecular electronic description and that the previously assumed hybridization of the σ frame had little resemblance to those correctly computed, and finally to show that some π-electron orbital energies can be at lower energy than some σ-electron orbital energies, with obvious consequence on the molecule's reactivity and spectroscopy. In the early paper (71) the bond's structure was, however, "drawn" using the traditional notations and with no direct derivation from the quantum chemical computation.

Today, however, we can say somewhat more about the chemical formulae and about the electronic distribution along the bonds. <u>First of all the formula is obtained from the D matrix.</u> As for the population in a bond, we can state that a single bond only seldom contains <u>exactly</u> two electrons, contrary to the classical proposal.

An aromatic bond, however (π bond) tends to be more confined in its charge distribution and contains nearly 1 π electron per bond. The mobility of the electrons, a traditional concept for π-bonds, seems to be not only a characteristic of π-bonds, but a <u>general</u> characteristic for most bonds.

Bonds are often asymmetrical in the distribution around the midpoint, one half being often more electronically crowded than the other half. As a corollary, the assumption of bond moments often made in empirical computations, especially in infrared studies, can be obtained from the analysis of ab initio wave functions and needs not to be an empirical concept.

Accumulation of data on several molecules will allow establishment of firm correlations between bond populations in different environments and bond populations versus bond energy of the BAP type.

It seems that we are not too many years away from the time when we shall relatively seldom need to perform ab initio computations in the way done today because proper tabulation of BAP type information and bond density will be available, thus allowing data to be obtained directly by interpolation of tables. Stated differently, the overall picture emerging from ab initio computation starts to be sufficiently complete to allow more correlations and less computations. With the molecular orbitals theory, the traditional concepts related to chemical formulae did at times appear as less and less essential; however, by extending the molecular orbital analysis, some of the traditional chemical bond concepts come back to us stronger because they are clearer and with it a deeper definition of the chemical formulae is available.

We expect that the lone pair concept will receive more attention as a prototype of a dynamic definition of bonds needed for reactivity. From Lewis' assignment of two-electrons per lone pair, we did pass to the MO representation, where we learn that the lone pair can be somewhat delocalized; but this is equivalent to state that along the lone pair direction, the electronic population does not need to be equal to two electrons (as is the case for most bonds). We propose to call the lone pair electrons, a virtual bond, because of its importance in reaction-chemistry and mainly because <u>it is</u> a bond. Essentially a bond is a localization of electronic charges often equally subdivided into two populations with opposing spins; the electronic population is characterized <u>in addition</u> by directionality. The traditional picture that we need <u>two nuclei</u> at its extreme is important but not essential and merely provides a useful and easy identification for the directionality (directionality does not necessarily imply two nuclei). In this sense the lone pair is a bond; a virtual bond for reactivity, but a real bond in its electronic structure. In this sense, some of the contradictions and ambiguities in describing inner shell electrons seems to have been overcome, since the electronic density associated to inner shell electrons lacks in a given space directionality, <u>one of the two essential ingredients of a bond</u>. The symmetry constraints that we can impose on the inner shell electrons reveal a somewhat "unphysical" character, since symmetry does not necessarily provide directionality, namely strong local anisotropy.

In conclusion, from the delocalized MOs (containing two electrons) we obtain a localized and directional distribution of electrons containing a variable number of electrons; often but not always the distribution is delimited by two nuclei. This picture seems to provide the flexibility needed to discuss reactivity and, at the same time is based on a natural extension of the classical concepts of chemical bond. Finally inner shell electrons and Rydberg electrons are now very naturally accounted for. Concerning the latter we can re-analyze the problem of chemical formulae in excited states, a somewhat neglected area. Again, the pioneering work of Mulliken in the Rydberg states, namely the MO picture, is the starting point from which one can then discuss chemical formulae for Rydberg states.

3.10 <u>BEA AND REACTION SURFACE</u>

One of the first steps in a theoretical study of a reaction is often the computation of an energy surface, since we postulate that generally the reaction will "follow" the "main valley" of such a surface. However, if the reaction involves two moderately large molecules, the energy surface requires a large number of computations that, if carried out ab initio, could easily go beyond the present practical computational feasibility. As known, for a system of m atoms the degrees of freedom are 3m-6 but to a first approximation and with chemical common sense, fewer need to be considered. Even so, often there are still too many "chemical degrees of freedom" which represent the parameters for the surface. For simplicity, let us assume a parabolic behavior for each parameter (about three computations are needed for each parameter) and further, let us assume a nearly linear independence of the parameters; still the total number of computations to be performed remains too large. This is the main reason why energy surfaces are either computed with semi-impirical methods or, in general,are available only for small systems, (when ab initio methods are used). It is, in addition, rather disappointing at the end of such computations to be limited by statements like "the computed barrier is XX kcal/mole" and to find no alternative either to analogies or to qualitative descriptions when one wishes to add even a limited amount of quantitative details on some chemical aspect of the reaction such as bonds rearrangements.

The two problems, namely the large number of computations and the limited amount of chemical information obtained, can be partially alleviated by performing a bond-energy analysis. Here, we shall present an example; we think that, even if the example is a very specific one, the conclusions are of interest for most studies of potential energy surfaces.

In Figure 8 we present a model for a catalytic site, a titanium compound, interacting with C_2H_4. The reaction we study is a typical Ziegler-Natta polymerization reaction; we are mainly interested in the initial part of the process, namely the insertion of the olefin. In our model, at the end of the reaction, we shall find two new bonds, Ti--C(4) and C(3)--C(5), one bond will be broken (the Ti--C(3) bond) and one is modified, the C(4)==C(5) double bond is transformed into the C(4)--C(5) single bond (see Figure 8 for the symbols of the atoms).

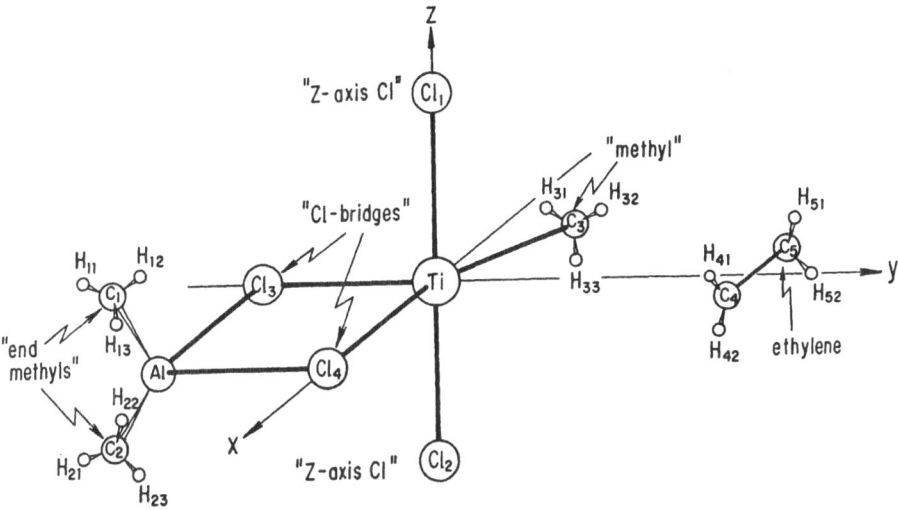

Figure 8. Ziegler-Natta reaction. Assumed model for the catalytic active site.

There are 24 atoms involved in this reaction, but since we shall consider the $Al(CH_3)_2$, the Ti atom and the four Cl atoms as fixed, the number of parameters is small, namely, the five distances Ti--C(4), Ti--C(5), Ti--C(3), C(2)--C(5), C(4)--C(5) and four angles α_T, α_R, α_O and α_I that correspond to 1) a rotation of CH_3 about the Ti--C(3) bond, 2) a rotation in the x, y plane of the Ti--C(3) bond about the fixed Ti atom, 3) the rotation of the hydrogen atoms of --CH_3 H(31), H(32),H(33) of Figure 8 about the Ti--C(3) bond, and 4) the internal angle H(51)--C(5)--H(52).

Arbitrarily, we have considered 14 steps in our analysis in order to describe the reaction from the initial configuration (C_2H_4 at about 10 A from the catalyst) to the final one (formation of the propylene chain). From our previous discussion we would require at least (9x3)x14=378 computations, more than we could financially afford. After few preliminary computations (73), intended to verify the catalyst geometry and to "feel" the surface, 12 steps were initially selected (74) by assuming concerted motions of the atoms, and later (75) two more were added.

The physical idea underlying a concerted motion is that each parameter plays its role in a sequential and ordered fashion, few parameters are dominant at each step, the remaining being of secondary importance. At the beginning, the dominant parameters are the distances, Ti--C(4) and Ti--C(5) and C(3)--C(5), since the ethylene molecule most come near to the model-catalyst from some assumed direction (we have analyzed the incoming of the olefin along the y axis, with the C_2H_4 molecular plane in the xy plane and with the C(4)=C(5) bond perpendicular to y axis). One expects that the CH_2 group of the olefin interacts in a repulsive way with the methyl group bound to Ti; as a consequence the methyl group tends to move away by "turning on" the α_R parameter. In this way in the intermediate stage of the complex formation, Ti--C_2H_4 is obtained. Then, the most important parameters are the rotation of CH_3 about the Ti--C(3) bond, the lengthening of the Ti--C(3) bond and the rotations of the hydrogen atoms of the CH_2 group with concomitant rehybridization from sp^2 to sp^3. At the same time, the carbon-carbon double bond must become a single bond (that is, the bond length increases from 1.33 to 1.54 A).

In Figure 9 (top) the computed total energy of the first attempt is presented (curve A). Two barriers were obtained, the first (step 6) with a barrier height of 17 kcal/mole, the second (step 9) with a barrier height of 15 kcal/mole. The results on the basis of experimental data could be considered as acceptable. The bond-energy analysis for each step is reported in Figure 9 (right insert) where we display only the interactions between the pair of atoms C(3)--C(5), Ti--C(4) and Ti--C(3). The diagrams indicate very clearly a bond formation (the bond energy from its near to zero original value, decreases to large negative values) for C(3)--C(5) from step 9 to step 12, preceded by a strong repulsion (from step 2 to step 8); a bond formation for Ti--C(4) from step 8 to step 12 preceded by some complex stabilization (minimum at step 6) and a bond breaking (bond energy coming to zero from a negative, namely, attractive value) for Ti--C(3), from step 7 to step 12. It is very interesting to note that the breaking of Ti--C(3) nicely compensates the energy of the Ti--C(4) formation; the reaction seems to be dominated by the interaction of the CH_3 and the CH_2 groups (of which we report the C(3)--C(5) pair). We note that none of the above comments could be made without a bond-energy analysis; the comparison of orbital by orbital density variations tells us of the flow of charges and only with some help from imagination and good sense one can hope to obtain a relationship between flow of charges and bond formation or breaking.

Since, a reaction "step" has little physical meaning unless it is performed in a time dependent framework, we have plotted the bond energies versus the bond inter-nuclear distances (see Figure 9, left insert). The C(3)--C(5) bond energy in Figure 9 (left inset) reveals a complex behavior with strong energy variations at distances between 2.2 and 2.8A. The arrows indicate the ordered sequence of the steps leading to the final state. Such energy variations are against the assumptions underlying the reason for a concerted motion, namely, as smooth a transition as possible; therefore we take such sharp variations in energy as an indication that the path is not optimized (75). For the formation of the Ti--C(4) bond and the breaking of the Ti--C(3) bond, the diagrams are reasonable and we take this as an indication (although not a proof) that the path is acceptable for such interactions. Thus, we decided to perform a few more computations attempting to decrease the C(3)--C(5) repulsion by shifting in a more careful way the methyl group during the C_2H_4 approach. The resulting total energy is indicated as B in Figure 9 (top): the first barrier no longer exists. The bond energy versus step number, as well as the bond energy versus bond distance diagrams now display a more reasonable aspect (see Figure 9; left insert, curve b; right insert, curve b). We stress that the choice of the important parameter to be varied was not easy on the basis of the information available from the total energy but was determined very simply by inspection from the bond energy analysis.

Finally, a few more parameters were optimized (75), in particular the C==C bond length that affects the Ti--C(3) and Ti--C(4) distances.

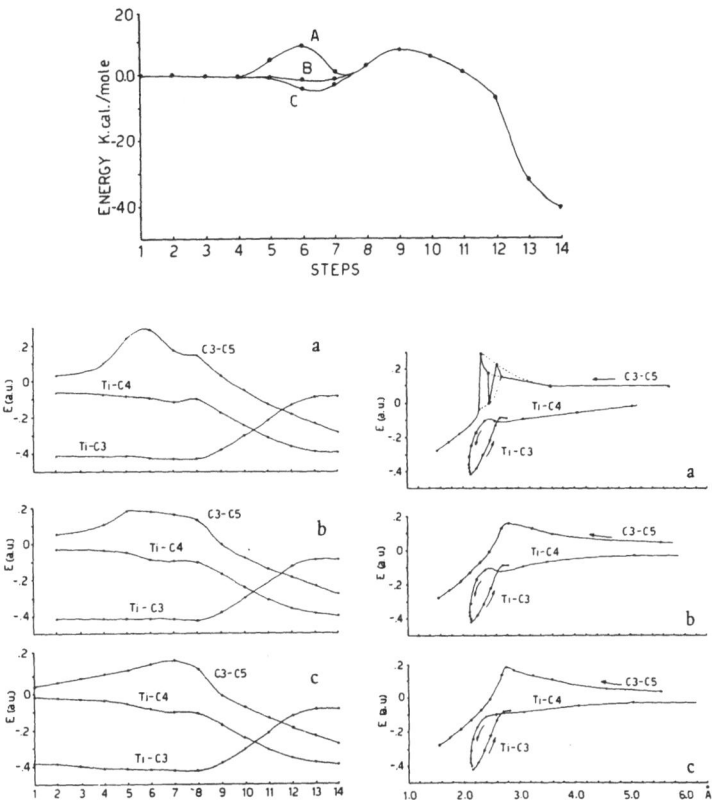

Figure 9. Top Insert. Total energy for the Ziegler-Natta reaction as obtained at the three different stages of the analyses, curves A and B refer to the first and second stage (reference 73 and 74); curve C refers to the last stage. Left Insert. BEA-bonds as obtained at the three different stages of the analyses. The curves of the inserts a, b, and c correspond to the total energy curves A, B, and C. Only three bonds are presented and the corresponding BEA-bond energies are given as function of the reaction step indices. Right Insert. BEA-bond energies for three bonds given as function of the bond length. The inserts a, b and c correspond to the three stages of the analyses of the reactions with the total energies A, B, and C. The arrows indicate the direction from initial to final products, thus follows the reaction step indices.

The results are those reported in Figure 8 (curve C) and in the bottom inserts of Figure 9 (left insert, curve c and right insert curve c). The complex "Ti-olefin" becomes even more stable and the nonphysical irregularities of the bond energy for Ti--C(3) and Ti--C(4) between 2.4 and 2.8A disappear, leading to the smooth partitioning given at the curves c, of the right and left inserts of Figure 9.

We call attention to the result that the bond energies corresponding to the Ti--C(3) and Ti--C(4) interactions do not seem to feel the drastic variations of C(3)--C(5) interaction in passing from the conformations corresponding to the top, the middle and the bottom inserts of Figure 9. This informs us of the local nature of bonds. Notice that this result is obtained by an analysis on a delocalized description of the molecule, namely the standard SCF-LCAO-MO functions.

We hope that this example has pointed out that the bond energy analysis, for example, the BEA-formalism, can be used to decrease the number of computations needed to scan the energy surface and to provide information not easily extracted from the total energy.

3.11 BAP AND REACTION SURFACE

In this section we shall use the BAP analysis, rather than BEA, to follow a reaction pathway. In the example we shall consider that the traditional bonds of the initial state are very different from those of the final state. Thus we are confronted with the problem of specifying which bonds should be considered. We recall that in BAP the non-bonded interactions are partitioned starting from the BEA-bond energies. In this example we have defined the BEA-bonds not automatically by program (by simply selecting a given energy threshold, standard BAP procedure), but by furnishing as input to the program a preassigned list of pairs of atoms consisting of the BAP for the initial step and of the BAP of the last step. In our example, we consider the reaction $H_2O + CO_2 \rightarrow H_2CO_3$ in presence of Zn^{++} a problem of interest in the enzymatic reaction with carbonic anhydrase (76,77).

We would like to stress that these computations are still preliminary and are presented not to discuss the mechanism of the reaction in the enzyme, but to discuss the BAP formalism. Indeed in these computations we have not included the reaction field of the amino acid residues at the enzyme's active site, a point of basic importance in any enzymatic reaction as previously discussed for the case of papaine interacting with the substrates.

First, we briefly outline the reaction pathway we have selected, performing a standard analysis limited to the total interaction energy. For details on basis set, geometries and over all discussion concerning the carbonic anhydrase enzyme we refer to (77).

From a previous study (76), we know that more than one water molecule can take part in the reaction at the enzyme active site. It is known in addition that any realistic mechanism of reaction would require the explicit inclusion of the field of Zn^{++} and of the amino-acid residues forming the cavity. For example, the field of the glutamate residue, if in the ionic form, can be expected to play an important role in the reaction. It is clear, however, that the field of the cation Zn^{++} is dominant in the carbonic anhydrase reaction. For this reason, in the following, we shall examine the formation of H_2CO_3 in the field of Zn^{++}. We assume that in the initial conditions, Zn^{++} is bound to at least one water molecule.

We have computed a number of reaction steps, that in turn can be grouped into a few chemical sequences. In the first sequence we assume that a CO_2 molecule approaches the $(Zn-H_2O)^{++}$ complex forming a precursory complex, where both $(Zn-H_2O)^{++}$ and CO_2 maintain their original geometry. Then, in the second sequence, a complex is formed between $Zn^{++}-H_2O$ and CO_2; the complex is obtained by concerted motions, whereby the O-C-O bond angle decreases from 180° to 125°, the C-O bonds lengthen, and the carbon atom of CO_2 forms a bond with the water oxygen. It is shown that the formation of this complex requires little energy: a small energy barrier of about 9 kcal/mole is found, but this value is only an upper limit in the enzymatic reaction since we have not included the reaction field effect.

The next step in the sequence is the removal of the complex by sandwiching a second molecule of H_2O between Zn^{++} and the complex. The

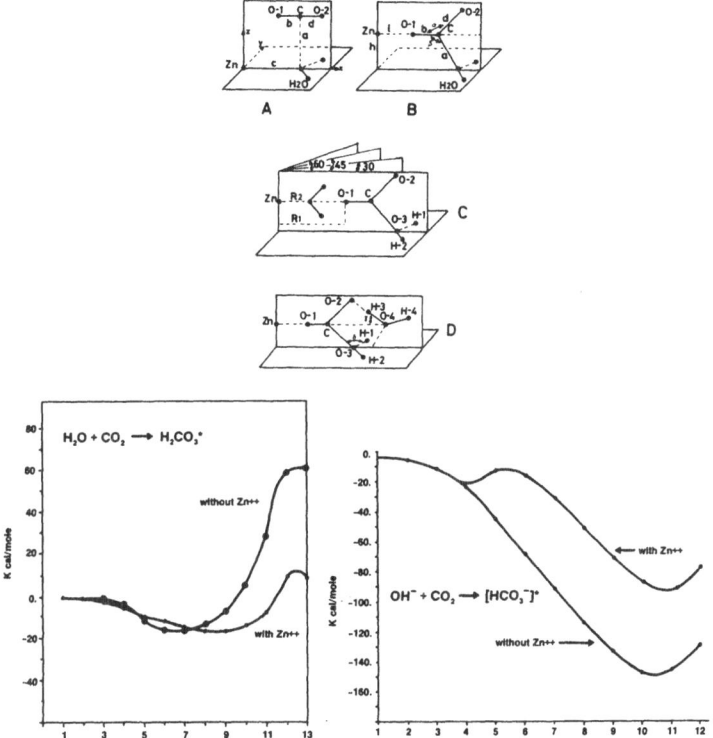

Figure 10. Top Insert. Main events in the reaction $Zn^{++}+CO_2+H_2O$. A refers to the precursory complex, B to the final H_2CO_3 complex, C to the removal of the complex by a second water molecule (which can lie in the planes at 0°, 30°, 45° or 60°) and D to proton transfer via a H_3O^+ intermediate. Left-Bottom Insert. Energy for the formation of the intermediate and final complex, as reported in Tables 16 and 18. Bottom-Right Insert. Total energy for the $OH^-+CO_2 \rightarrow (HCO_3)^-$ reaction in presence or in absence of Zn^{++}, as reported in Tables 20 and 22.

last step of the sequence is a deprotonation at one O-H bond and protonation of an oxygen atom in CO_2. This can easily be obtained by intervention of a third water molecule yielding a hydrated carbonic acid. The last two sequences can be either inverted in the order or considered to occur as a concerted event. These two steps are analyzed elsewhere (77). The details of this sequence of events are described by the reaction steps reported in Table 16 and in Figure 10. In this figure we report the 4 main events for a reaction mechanism, its total energy (including or excluding Zn^{++}) the total energy for the equivalent section steps considering OH^- rather than H_2O, either in presence or in absence of Zn^{++}. In steps 1 to 4 the first event, the formation of a precursory

Table 16. Energies and geometrical parameters for the reaction mechanism H_2O+CO_2 in presence of Zn^{++}.

No.	R(O3-Zn) c	R(C-O3) a	R(O1-C) b	R(O2-C) d	O1-C-O2	O1-C-O3	Zn-O1-C	R(Zn-O1) e	E (a.u.)	ΔE (kcal/mole)
1	3.60	8.0	2.2014	2.2014	180	90	100	8.12	-2031.971271	-74.0
2	3.60	7.0	2.2014	2.2014	180	90	101	7.14	-2031.972181	-74.6
3	3.60	6.0	2.2014	2.2014	180	90	103	6.16	-2031.974614	-76.1
4	3.60	5.0	2.2014	2.2014	180	90	106	5.19	-2031.978295	-78.4
5	4.08	4.64	2.2389	2.2428	174	93	110	4.68	-2031.986593	-83.6
6	4.57	4.28	2.2764	2.2641	168	95	118	4.20	-2031.988988	-85.1
7	5.07	3.94	2.3139	2.3255	162	98	125	4.46	-2031.993402	-87.9
8	5.58	3.61	2.3519	2.3668	156	102	134	3.47	-2031.996262	-89.7
9	6.09	3.30	2.3889	2.4082	148	106	145	3.25	-2031.994235	-88.4
10	6.09	3.01	2.4264	2.4495	140	111	157	3.16	-2031.991319	-86.6
11	7.12	2.75	2.2464	2.4909	132	118	178	3.21	-2031.984186	-82.1
12	7.64	2.53	2.5014	2.5322	125	125	180	3.40	-2031.952495	-62.2
13	7.64	2.53	2.5014	2.5333	125	125	180	3.40	-2031.957148	-65.1

Note: In all steps the OH distance is 1.8088 a.u. (exception for STP 13 where OH=2.3088 a.u.)

complex, is considered; the binding energy of this complex is only -4.4 kcal/mole as indicated in Table 16; the geometry of the complex is illustrated in the insert A of Figure 10. The energy is reported relative to the three noninteracting species: H_2O, CO_2 and Zn^{++}.

In steps 5 to 13 (the second event) the formation of a complex between H_2O and CO_2 in the field of Zn^{++} is considered; the final configuration of the complex is represented in the insert B of Figure 10. The energetic details of the formation of this complex, given in steps 6 to 13 of Table 16, show that we propose a concerted motion, whereby both C==O bonds of CO_2 length and at the same time the carbon atom rehybridizes

Table 17. BAP for selected pairs in the $CO_2+H_2O \longrightarrow H_2CO_3$ reaction (in a.u.) in presence of Zn^{++}.

step	Zn-O(1)	Zn-O(2)	Zn-O(3)	O(3)-H(1)	O(3)-H(2)	O(1)-C	O(2)-C	O(3)-C
1	-0.016	-0.010	-0.071	0..066	-0.066	-0.037	-0.035	-0.002
2	-0.018	-0.010	-0.069	-0.064	-0.064	-0.039	-0.036	-0.002
3	-0.021	-0.009	-0.067	-0.063	-0.063	-0.041	-0.037	-0.003
4	-0.027	-0.009	-0.063	-0.061	-0.061	-0.045	-0.040	-0.004
5	-0.032	-0.007	-0.058	-0.061	-0.061	-0.049	-0.042	-0.004
6	-0.040	-0.006	-0.049	-0.060	-0.060	-0.053	-0.045	-0.006
7	-0.047	-0.006	-0.041	-0.058	-0.058	-0.057	-0.048	-0.008
8	-0.051	-0.006	-0.034	-0.056	-0.056	-0.058	-0.052	-0.013
9	-0.050	-0.007	-0.029	-0.053	-0.053	-0.058	-0.055	-0.019
10	-0.049	-0.007	-0.024	-0.051	-0.051	-0.056	-0.057	-0.025
11	-0.050	-0.008	-0.020	-0.048	-0.048	-0.053	-0.057	-0.029
12	-0.046	-0.008	-0.015	-0.043	-0.043	-0.047	-0.052	-0.028
13	-0.050	-0.009	-0.014	-0.039	-0.043	-0.050	-0.055	-0.028

from sp to sp^2, binding to the oxygen of water. The final complex is formed with an energy expense of 1.8 kcal/mole. We stress, however,

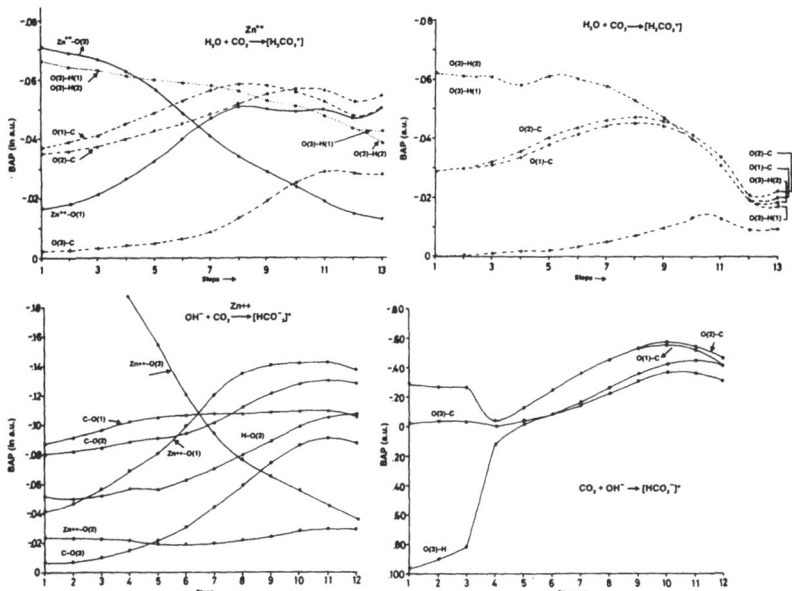

Figure 11. **Top Left Insert.** Decomposition of the total interaction energy into BAP contributions in the $H_2O+CO_2 \rightarrow (H_2CO_3)^*$ reaction in presence of ZN^{++}. (see Table 17). **Top Right Insert.** Bond Pair Analysis BAP of the reaction $H_2O+CO_2 \rightarrow (H_2CO_3)^*$ reaction in absence of Zn^{++} (see Table 19). **Bottom Left Insert.** Bond Pair Analysis for the $OH^- + CO_2 \rightarrow (HCO_3)^-$ reaction in presence of Zn^{++} (see Table 21). **Bottom Right Insert.** Bond Pair Analysis for the $OH^- + CO_2 \rightarrow (HCO_3)^-$ reaction in absence of Zn^{++} (see Table 23).

that the geometry of this complex has not been optimized. For example, an increase in the bond length of H(1) by only 0.5 a.u. lowers the barrier to 8.9 kcal/mole (lengthening of this bond lowers the repulsion between C and H, and stabilizes the second OH bond). Further geometry optimization in particular in the three O-C-O bond angles (assumed as 125°, 125°, 110°) and in the position of Zn^{++} will reduce this barrier even further. Indeed a proper optimization of the geometry of this complex could even make the barrier disappear; however in this work our main interest is to show that this complex can be formed at little energy expense. Let us now return to BAP. The list of atom pairs we have selected includes the traditional bonds for CO_2 and H_2O and the expected strong interactions between Zn^{++} and the oxygen atoms namely O(1) and O(2) for CO_2 and O(3) for H_2O. Thus the total list is Zn-O(1), Zn-O(2), Zn-O(3), O(1)-C, O(2)-C, O(3)-H(1), O(3)-H(2) and O(3)-C needed when the complex $CO_2 \cdot H_2O \cdot Zn^{++}$ is formed.

In Table 17 and in Figure 11 (top left insert), the corresponding BAP energies are given for the 13 steps of the reaction pathway. In Figure 11 we have omitted the BAP for the Zn^{++}-O(2) pair, since nearly constant during the entire reaction process (from step 1 to step 13). During complex formation the O-H bonds (namely O(3)-H(1) and O(3)-H(2) loose energy, whereas the C-O bonds (C(3)-O(1) and C(3)-O(3)) gain energy, in different amount since the field of Zn^{++} and H_2O are not symmetrical relative to the two C-O bonds. The main energy variations during the reaction are associated with three pairs: the O(3)-C that is nearly not interacting at step 1 and becomes a bond at step 13, the

Table 18. Total energy (in a.u.) and interaction energy (in Kcal/mole) for the 13 steps of the reaction $H_2O + CO_2$ without Zn^{++}.

Step	T.E.(a.u.)	ΔE
1	-262.676718	-0.23
2	-262.677194	-0.53
3	-262.678270	-1.20
4	-262.679002	-1.66
5	-262.693356	-10.67
6	-262.701485	-15.77
7	-262.703741	-17.18
8	-262.699995	-14.84
9	-262.688525	-7.64
10	-262.667999	5.24
11	-262.633688	26.77
12	-262.580839	59.93
13	-262.578770	61.23

$E(CO_2) = -186.933655$; $E(H_2O) = -75.742696$

Table 19. BAP for slected pairs in the $CO_2+H_2O \longrightarrow H_2CO_3$ reaction (in a.u.) without Zn^{++}.

step	O(3)-H(1)	O(3)-H(2)	O(1)-C	O(2)-C	O(3)-C
1	-0.062	-0.062	-0.029	-0.029	0.000
2	-0.061	-0.061	-0.030	-0.030	0.000
3	-0.061	-0.061	-0.031	-0.032	-0.001
4	-0.058	-0.058	-0.033	-0.035	-0.002
5	-0.061	-0.061	-0.038	-0.040	-0.002
6	-0.060	-0.060	-0.041	-0.043	-0.003
7	-0.058	-0.058	-0.044	-0.046	-0.005
8	-0.053	-0.053	-0.045	-0.047	-0.007
9	-0.047	-0.047	-0.044	-0.046	-0.010
10	-0.040	-0.040	-0.040	-0.041	-0.013
11	-0.031	-0.031	-0.031	-0.034	-0.013
12	-0.019	-0.019	-0.019	-0.021	-0.009
13	-0.017	-0.018	-0.019	-0.022	-0.009

Table 20. Total energy (in a.u.) and interaction energy (in Kcal/mole) for the 13 steps of the reaction $OH^- + CO_2$ with Zn^{++}.

Step	T.E.(a.u)	ΔE
1	-2031.692051	-350.38
2	-2031.694682	-352.03
3	-2031.701310	-356.19
4	-2031.716673	-365.83
5	-2031.704888	-358.43
6	-2031.708331	-360.60
7	-2031.732911	-376.02
8	-2031.766026	-396.80
9	-2031.797502	-416.55
10	-2031.820697	-431.10
11	-2031.830578	-437.31
12	-2031.806346	-422.10
13	equal to step 12	

$E(CO_2) = -186.933655$; $E(OH^-) = -75.023025$; $E(Zn^{++}) = -1769.1770$

Table 21. BAP for selected pairs in the $CO_2+OH^- \longrightarrow HCO_3^-$ reaction (in a.u.) with Zn^{++}.

step	O(3)-H	Zn-O(3)	C-O(3)	Zn-O(1)	Zn-O(1)	C-O(1)	C-O(2)
1	-0.052	-0.226	-0.007	-0.041	-0.024	-0.088	-0.081
2	-0.051	-0.217	-0.008	-0.047	-0.024	-0.092	-0.083
3	-0.052	-0.206	-0.010	-0.056	-0.023	-0.097	-0.085
4	-0.057	-0.188	-0.016	-0.069	-0.022	-0.103	-0.089
5	-0.057	-0.155	-0.022	-0.081	-0.019	-0.105	-0.091
6	-0.063	-0.121	-0.032	-0.100	-0.019	-0.106	-0.094
7	-0.071	-0.094	-0.044	-0.121	-0.020	-0.108	-0.102
8	-0.080	-0.076	-0.059	-0.136	-0.022	-0.108	-0.112
9	-0.090	-0.065	-0.074	-0.141	-0.025	-0.109	-0.121
10	-0.099	-0.056	-0.086	-0.142	-0.028	-0.109	-0.128
11	-0.106	-0.047	-0.092	-0.143	-0.029	-0.110	-0.131
12	-0.107	-0.037	-0.088	-0.137	-0.030	-0.106	-0.128
13	equal to step 12						

Table 22. Total energy (in a.u.) and interaction energy (in Kcal/mole) for the 13 steps of the reaction $OH^- + CO_2$ without Zn^{++}.

step	T.E.(a.u.)	ΔE
1	-261.961466	-3.00
2	-261.964844	-5.12
3	-261.973801	-10.74
4	-261.992260	-22.33
5	-262.027665	-44.54
6	-262.064222	-67.48
7	-262.101299	-90.75
8	-262.136693	-112.96
9	-262.169102	-133.30
10	-262.189159	-145.88
11	-262.188203	-145.28
12	-262.158510	-126.65
13	equal to step 12	

$E(CO_2) = -186.93365$; $E(OH^-) = -75.023025$

Table 23. BAP for selected pairs in the $CO_2 + OH^- \rightarrow HCO_3$ reaction (in a.u.) without Zn^{++}.

Step	O(3)-H	O(1)-C	O(2-6)	O(3)-C
1	0.096	-0.029	-0.029	-0.003
2	0.090	-0.027	-0.027	-0.004
3	0.081	-0.027	-0.027	-0.004
4	0.012	-0.003	-0.003	-0.001
5	-0.002	-0.013	-0.013	-0.004
6	-0.009	-0.025	-0.025	-0.009
7	-0.017	-0.036	-0.031	-0.015
8	-0.027	-0.045	-0.045	-0.023
9	-0.036	-0.052	-0.053	-0.031
10	-0.043	-0.055	-0.057	-0.037
11	-0.046	-0.053	-0.055	-0.031
12	-0.042	-0.043	-0.047	-0.031
13	equal to step 12			

Zn^{++}-O(3) pair that is essentially a strong ionic bond till it is broken at step 13, and the Zn^{++}-O(1) pair that start as a weekly attracting interaction (step 1) and end as a strong ionic bond at step 13. During the reaction the energy associated to Zn^{++}-O(3) is replaced by the energy associated to Zn^{++}-O(1). It is noted that a small variation in the geometrical parameters of Zn^{++} relative to the rest of the complex can bring about a large energy variation; the small barrier in Figure 10 (bottom left insert) could likely turn out to be unreal if a more careful geometry optimization would be performed at each step.

In Figure 10 (bottom left insert) (and in Tables 18 and 19) we report the same 13 steps, this time in absence of the Zn^{++} field. In Table 18 we report the total energy and in Table 19 we report the BAP decomposition. By comparing the data of Figure 11 (top left and top right inserts), one obtains a clear visualization of the effect of Zn^{++} during the reaction. We learn that the Zn^{++} field increases the O-H bond strength and that the two C-O bonds in presence of Zn^{++} gain strengths in that part of the reaction that brings about re-hybridization from sp to nearly sp^2 (step 1 to step 8) and that the energy loss (from step 9 to step 13) is mitigated by the presence of Zn^{++}. (The small barrier of about 15 kcal/mole would increase to about 60 kcal/mole, if Zn^{++} were absent.)

As known, in the carbonic anhydrase reaction one can either postulate the formation of H_2CO_3 from CO_2 and H_2O, or of HCO_3^- from $CO_2 + OH^-$. The two different pathways are originated from assumption concerning charged side-chains in the active site near to Zn^{++} (78) and proton transfers. From an heuristic point of view, we can perform (at nearly zero computer time) an analysis on the energetics when we substitute

H_2O with OH^-, leaving all geometries in the 13 steps unchanged (by merely suppressing an hydrogen atom (in H_2O) and replacing it with an electron). As previously done we shall consider the Zn^{++} field either on or off. In Figure 10 (bottom left insert) we report the total energies (both with and without Zn^{++}), in Figure 11 (bottom left and right inserts) the BAP again with or without Zn^{++}. The corresponding numerical values are reported in Tables 20 and 21 (total energy and BAP, respectively) in presence of Zn^{++} and in Tables 22 and 23 (total energy and BAP, respectively) in absence of Zn^{++}. Energetically (total energy) the reaction occurs without energy barriers. We note, however, that this result should not necessarily be related to the carbonic anhydrase since the reaction field of the enzyme has not been included in the computations.

As in the previous example for the BEA, here again we are not too interested to discuss a specific chemical reaction but rather to indicate the versatility of BEA and BAP to discuss reaction pathways. Work is in progress to study the reaction mechanism of this enzyme.

3.12 BOND ENERGY ANALYSIS AND VIBRATIONAL ANALYSIS

As pointed out in the first work concerning BEA (54), it seems that any vibrational analysis could gain if performed not on the total energy but on the bond energies. To start with it makes little sense to differentiate the total energy, especially for large molecules, when only part of the total energy is related to a given vibration. Secondly, the selection of the vibrations that are coupled one to another is often rather arbitrary, whereas it becomes immediate using BEA-bonds. Thirdly, the form of the potential can be properly selected either for attractive or for repulsive interactions and equivalently for coupling terms between vibrations. Fourthly, the bond-dipole nature of bonds can be included in a consistent way. Finally, the validity and the limit of concepts, such as group frequency, emerge very naturally from the BEA representation; perturbation induced by other molecules (on the one analyzed) is also most easily visualized and included by performing the vibrational analysis on the BEA-bond energies rather than on the total energy.

4.0 ATOM-ATOM PAIR POTENTIALS

4.1 PRELIMINARY COMMENTS

When the complexity in a chemical system results from the number of molecules or atoms in the system, the determination of the intermolecular interactions becomes of dominant interest. This point has been fully realized since the beginning of quantum chemistry by those researchers who were interested in transport phenomena in the liquid state and, in general, in thermodynamics.

In the decade of 1960, however, the main effort of quantum chemistry was directed at attempting to obtain exact solutions to the Schroedinger equation. This caused a concentration of efforts on atomic and small molecular systems.

As known, given a system of N molecules, A,B,C,...N, composed of atoms $1(A),...,i(A),...N(A); 1(B),...j(B),...N(B);...$ its total interaction can be written as:

$$V = \Sigma V(i,j) + \Sigma V(i,j,k) + ... + V[1(A),..., N(N)] \qquad 4.1$$

where the first term is the two-body term, the second is the three-body term, and the last term contains all the bodies of the system. The assumption $V = \Sigma V(i,j)$ is known as pair-wise additive approximation; in this contest the remaining terms of the series are often referred to as "nonadditive corrections." At the base of this expansion there is the underlaying hypotheses that there are entities recognizable as atoms in a molecule. Roughly, we assume, as apparent from example from an X-ray defraction pattern, that we can associate a given set of electrons to a given nucleus, and refer to this combination as an "atom." On the other hand in quantum chemistry we have nuclei and electrons, and the hamiltonian does not recognize the existence of "atoms" in a molecule. A quantum mechanical definition for atoms and molecules has been previously discussed to overcome this semantic problem.

4.2 ATOMIC CLASSES FOR ATOMS IN MOLECULES

In the Heitler-London-Stater-Pauling valence bond approximation, one finds the familiar concept of valence state, namely a hypothetical atomic state, with space orbitals optimally hybridized and therefore pre-arranged for the electronic structure of that atom when inserted in a molecule (59,62). To obtain such electronic distribution there is an energy loss relative to the ground state distribution of the atom. Previously, we have introduced the molecular orbital valence state (MOVS) concept that parallels in the MO description what was done in the valence-bond description. The MOVS, however, is obtained after one has the molecular wave function, not independently from it. As a consequence not only hybridization, but also charge transfer is accounted for and therefore either an energy loss or an energy gain accompanies the redistribution. The MOVS represents a first characterization of an atom when in a molecule. A second characterization is the gross charge (or the net charge) partitioning of the molecular electronic density as proposed by Mulliken in his electron population analyses (1). We refer to the third chapter for more details. In the following we shall stress the implications and applications of the concepts previously described.

Let us, as an example, consider the MOVS energies and the net charges (NC) for the oxygen atoms in two familiar classes of compounds, namely the naturally occurring amino-acids in the neutral form, and the four

bases in DNA. From hystograms for such quantities, we can notice three
main distributions, one associated with the oxygen atom of C=O in COOH,
a second one with the oxygen atoms of C-O-H in COOH, and a third one
associated with the C=O of the bases. The three MOVS energies have
the values of 0.61 ± 0.03 a.u., 0.38 ± 0.03 a.u. and 0.33 ± 0.02 a.u.;
for the three cases the net charges value are -0.53 ± 0.06 electrons,
-0.40 ± 0.05 electrons and -0.39 ± 0.01 electrons for the oxygen of
COH in COOH, CO in COOH and CO in the bases. We can note that it takes
more energy to perturb an oxygen atom from the ^3P ground state into the
optimal form for COH than for CO (in COOH); this is equivalent to say
that we perturb three rather than two electrons, as known from the
Lewis-Langmuir representations

: C::O| and :C:Ō:H

The complementary representation, via population analyses (see the
above values of the net charges), is that charge transfer is larger in
COH than in CO for COOH. These two different situations for the oxygen
atoms can be stated by asserting that the oxygen atom can be in two
different "classes," a class representing a specific electronic environment probed by the MOVS energy and the net charge values. Originally
each class was designated by a code number starting from one and increasing progressively with the growth of our library of potentials.
More recently, we have used both the atomic symbol and a code number
(75). For example the oxygen atom COH in the COOH did correspond, in
the early notation, to class number 9, the oxygen atom of CO in COOH
to class number 10, the oxygen atom of CO in the bases to class number
27; in the new notation these classes are designated as O-03, O-06 and
O-11, respectively (79,80,81,82). Let us now consider a more general
example designed to show how to subdivide atoms into classes.

In the amino acids and in the four bases of the nucleic acids the oxygen
atoms cluster into very few classes. A more complex situation is offered
by an analyses of the carbon atoms in the same set of molecules. A
class definition collects the following number of information: the Z
value of the atom, the number NL of atoms that are bonded to it, according to the BAP analyses, and finally the range of the values for the
net charges, NC, and MOVS energies characteristic for that class. The
latter are supplied by giving the average value and the variance range;
for example a class can have an average value of NC = -0.12 electrons
and a range of ±0.02 and an average value of MOVS = 1.08 a.u. with a
range ±0.02. For convenience, often we supplied the list of the atoms
1(1),1(2),... that are bound to the atom in consideration. The full
class is therefore characterized by a class code, namely an identification (CLASS ID) and by the following parameters.

| CLASS ID | Z | NL | MOVS | Δ | NC | Δ | 1(1) | 1(2) | 1(3) | ... |

For example, the two following environments for carbon atoms C*

```
      H                  N
      |                  |
      C*              C-C*-H
     / \                 |
    N   C                H
```

might have the following characterizations

CLASS ID	6	3	1.08	±0.02	-0.12	±0.02	N	C	H	
CLASS ID	6	4	1.06	±0.03	-0.30	±0.03	N	C	H	H

a class code, a Z value for C* Z=6, either 3 or 4 ligands, the MOVS
energy in a.u., the variance for the MOVS of that class, the NC in
electrons, the variance for the NC in that class, plus optional information such as the specifications bonded to C*. The list of the optional information can be a multiple list; for example, if the following
C* carbon atoms are in the same class

```
      C              C              H
      |              |              |
  C—C*—C         N—C*—C         O—C*—C
      |              |              |
      C              H              H
```

we write

CLASS ID	6	4	1.03	±0.04	-0.17	±0.05	C	C	C	C
							N	C	C	H
							O	C	H	H

With this in mind we can now proceed to give the general rules to define
when an atom belongs to a given class. Let us indicate with ΔNC the
range (from the maximum value to the minimum value) of the NC distribution for all the atoms of given atomic number Z, with $\Delta MOVS$ the range
of the MOVS distribution for the same atoms, with NC a fraction of ΔNC
such that $\Delta NC = NC \times N(NC)$, with MOVS a fraction of $\Delta MOVS$ such that
$\Delta MOVS = MOVS \times N(MOVS)$. Let us, in addition, define as the "NC, MOVS
plane," a plane that has the NC values as abscissa and the MOVS values
as ordinate. Finally let us indicate with $NL(i)$ the number of possible
bonds for the atom in consideration (in the above example there are 4
atoms connected to C*, therefore $NL(i) = 4$ and $i = 1$). The $N(NC)$ intervals (segmenting ΔNC) and the $N(MOVS)$ intervals (segmenting $\Delta MOVS$)
define $N(NC) \times N(MOVS)$ areas in the NC, MOVS plane, and each area defines a class; namely, all the atoms considered correspond to points
in the NC, MOVS plane, all points which falls into a given area NC x
MOVS have the same $NL(i)$ value belong to the same class.

The remaining problem is related to a choice for ΔNC and $\Delta MOVS$. Reasonable values for ΔNC ranges are from ±0.02 to ±0.07; reasonable values for
$\Delta MOVS$ are from ±0.02 a.u. to ±0.05 a.u. In the class definition given
above there is no need for "chemical good sense;" all one needs is a
starting geometrical definition for the nuclei of the systems, the
number of the electrons in the systems, an approximate solution of the
Schroedinger equation and the BAP analyses. The traditional classifications, like aliphatic carbon, aromatic carbon, carbonilic carbon, etc.,
are all contained in the "class" subdivisions. It is noted, however,
that we do not have available the NC and MOVS value for all the carbon
atoms, but only for those present in a limited sample of molecules; as
a consequence the $N(NC)$ and $N(MOVS)$ definition is partly based on chemical good sense.

In conclusion, an atom in a molecule can be characterized first by its
atomic number (Mendeleyev), then by its net charge (Mulliken) and finally by its MOVS energy (Clementi). The distribution of the atomic
numbers is discrete; the distribution of all possible net charges and
MOVS energy is, in general, a continuous distribution defined in a
given interval; the class concept segments the continuous distribution
into subintervals each one characteristic for a given electronic environment. The deviations from the average value of each subinterval
(each class) either in the net charge or in MOVS energy are the response
to nearest-nearest neighbors interactions. As previously explained, the
gross population and the MOVS energy are essential to determine the

chemical structural formulae and to define bonding and nonbonding interactions as well to quantum mechanically define an atom or a molecule in a system composed both by atoms and molecules. We have, therefore, a unified framework of concepts sufficiently extended to describe either simple or complex chemical systems.

4.3 DETERMINATION OF TWO-BODY PAIR POTENTIALS

In this section we shall consider the problem of the derivation of atom-atom pair potentials to describe intermolecular interactions. A first problem concerns the analytical form to be selected. Given two atoms i and j, belonging to the molecules P and Q, respectively, the interaction potential $V(i,j)$ might contain explicitly an angular and a radial dependency. Often in our work, we have selected to use only some function of the internuclear distance. This decision has been based on the fact that often we were not too interested in obtaining the exact partitioning of the interactions between the atom i in P and atoms j in Q, but rather to obtain a reasonable total interaction energy between P and Q. Since the atom i belongs to some class, which reflects the specific relations to its neighbors (and equivalently for j), the field due to i is also a function of its neighbors. As a consequence we attemp, via atom-atom pair potentials, to describe the intermolecular interactions between the atom i in its environment and the atom j in its environment; for this scope radial functions are sufficient. A well known equivalent situation (in quantum mechanics) is the representation of p,d,f... orbitals by a fixed linear combination of s orbitals; the equivalent to the fixed linear combination is the geometrical relationship of the atoms i and j to their neighboring atoms.

The next problems concern the form of the radial function describing the intermolecular interaction between the atoms i and j. The solution in our case represents a compromise between two conflicting requirements. For accuracy and flexibility (and from analogy to the multipole expansions) one would like to use a rather long series of terms in $r(i,j)$; on the other hand the larger the number of terms, the larger the number of associated fitting constants. We have often settled (with exception of particularly accurate potentials) for a three term expansion of the form

$$V(i,j) = B(a,b;i,j)/r(i,j)^{12} - A(a,b;i,j)/r(i,j)^{6} + \qquad 4.2$$
$$+ C(a,b;i,j)q(i)q(j)/r(i,j)$$

where A, B, and C are fitting constants for the atom i of class a and charge $q(i)$ and for the atom j of class b and charge $q(j)$. The quantity C should not be interpreted as related to the dielectric constant; it represents an average correction for the net charges $q(i)$ and $q(j)$ that are obtained from computations on the separated molecules P and Q. The net charges are a function of the internuclear separation and of the class, namely $q'(i) = f(q(i),r(i,j),a)$ and $q'(j) = f(q(j),r(i,j),b)$ such that for $r(i,j)$ equal to infinity we have $q'(i) = q(i)$ and $q'(j) = q(j)$. For $r(i,j)$ finite, we approximate $q'(i)q'(j)$ to $C(a,b;i,j)q(i)q(j)$. In the same way, the constants A and B are not to be equated with the formally similar Lennard-Jones parameters. The set of constants A, B, and C are mutually dependent but these constants are also dependent on the values of A, B, and C of the neighboring atoms to i and j; the dependency of $V(i,j)$ on the classes a and b allows the above drastic simplification.

In this sense the simple looking expression above given for the interaction represents a more complex expression that includes nearest neighbors to i and nearest neighbors to j.

A third problem, concerning pair potentials, is related to its transferability: from what has been previously explained, it is clear that the transferability of V(i,j) exists generally when the neighbors of i and of j in some molecules P' and Q' are essentially the same as the neighbors of i and j for the model molecules P and Q for which A, B, and C have been computed. In a strict sense, transferability occurs only for atoms belonging to the same class. Our definitions of atom-atom pair potentials as functions of the classes has, therefore, transferability built into it by explicit design and is not limited to a given family of molecules.

A final question concerns the original data from which atom pair potentials can be derived. Essentially this question is whether the potential should be derived from experimental data or from computational data. The answer depends on the availability of a sufficient number of comparable data from which the constants A, B, and C can be fitted. Let us consider as an example arginine interacting with a water molecule: the twenty seven atoms of arginine can be grouped into ten classes, the three atoms of water can be grouped into two classes, thus we have to extract from either experimental or computational data twenty potentials of the previously analyzed form, that is, we have to fit sixty constants (A, B, and C). As known, using fitting techniques, it is advisable to have several points, say n, for the determination of one fitting constant: for n = 4, we need 240 independent experimental or computational data. As known there are nearly no experimental data for our example. On the basis of this example it should be clear, why we have resorted to determine the fitting constants from computations rather than from experimental data. In the past five years we have determined the interaction potentials for twenty-six classes concerning hydrogen atoms, for thirty-three classes concerning carbon atoms, nine classes for the nitrogen atoms, two for sulphur and one for phosphorous atoms.

Since these classes have been used to study the interaction of molecules with water (two classes), we had to construct a total of 71 x 2 potentials, thus, over four hundred fitting constants have been determined. To obtain a reliable fitting, we have computed by ab-initio methods the interaction energy of about 4000 complexes. These interaction energies have been computed using a minimal gaussian basis set of 7/3 type for C,N,O: 9/5 or 9/6 type for S and P and 3/0 for H (79-82).

Recently, we have extended our library of potential by including not only the interactions for the atoms of the previous classes with a water molecule, but also with the Na^+ and Li^+ ion (83,84). These potentials are supplemented with the accurate one for ions with water (68,85,86,87) and water with water (88,89,90). In addition, another (and larger) set of pair potentials, namely those corresponding to the classes for the amino acids interacting among themselves, rather than with a water molecule or with ions (91). As a consequence the interaction potentials for example of any amino acid with another amino acid is available in a pair potential form. In this way we have computed 1218 constants A, B, and C; since the computational effort to compute several thousand ab-initio interactions between two amino acids is too costly for us, the A, B, and C constants are obtained, not only from the total interaction energy derived from ab-initio computations, but also from the bond energy analses. In this way the computational effort is reduced by a large factor and less than 2000 ab-initio computations for an amino acid - amino acid pair were needed.

4.4 PAIR POTENTIALS AND AB-INITIO COMPUTATIONS

The interaction energy from atom-atom pair potentials for two molecules

P and Q is given by the relation

$$I(P,Q) = \sum_{i,j} V(i,j) \qquad 4.3$$

whereas from ab-initio computations is given by the relation involving the energies

$$I(P,Q) = E(P,Q) - E(Q) - E(P) \qquad 4.4$$

As noted, the pair potentials are transferable as long as the molecules P and Q are composed only of atoms belonging to classes for which there are the corresponding $V(i,j)$.

With this in mind, we can discuss the reliability of the pair potentials. For an exact fitting the intermolecular interactions computed with pair potential are equal to the intermolecular interaction computed by ab-initio methods.

The gain in computational speed is too obvious to be discussed (several thousand times faster). In practice, however, any fitting invariably is associated to an error (standard deviation): for most of the cases we have reported in literature, the standard deviation varies between 0.5 and 2.0 kcal/mole, and corresponds to about 5% or 10% of the ab initio intermolecular interaction energy at the minima; details on the standard deviation are reported elsewhere (79-84) in the papers dealing with the construction of the $V(i,j)$ for the different model compounds.

We note that the use of pair potentials represents notable step forward over the use of the electrostatic potentials. In particular the pair potentials are valid for intermediate and short distances, whereas the classical electrostatic model either starts to fail or totally fails in such regions. In addition, the electrostatic potential is the response to a hypothetical probe, a point charge, rather than the response to a specific molecule. The attractive feature of the electrostatic potential is the ease in obtaining electrostatic potential maps (rather than the iso-energy maps) which might provide a general, rather than a specific, picture on reactivity sites for a molecule. The negative feature concerning the electrostatic potential maps, is the current abuse of the method, especially in "quantum pharmacology and quantum biology." The abuse is at two levels: 1) use of the electrostatic potential in regions where the approximation breakdown; 2) extrapolations much beyond the intrinsic validity limit of the approximation.

4.5 MINIMAL BASIS SET AND BASIS SET SUPERPOSITION ERROR

It appears, from the preceding sections, that a weak point in our scheme is not the use of pair potentials to obtain $I(P,Q)$, but rather in the deficiencies of quantum-mechanical computations, if carried out uncritically. As known, these suffer from at least two main errors, namely, basis set truncation and neglect of electron correlation corrections. The former error can be eliminated by properly extending the basis set orbital exponents at different values of $r(i,j)$: the latter can be eliminated by introducing full C.I. As known, neither of the two above suggestions has much practical value because of computer time cost. In the first chapter we dealt with techniques that should provide practical alternatives. In the following of this chapter, we shall analyze the problem of obtaining reliable inter-molecular interaction, but we shall neglect the problem of correcting the intramolecular interactions.

With today's standard techniques in dealing with two interacting molecules (of not trival size), one is compelled to use minimal basis sets. Let us consider this point in more detail. The term "minimal basis set" was introduced in the early quantum chemical literature (1960) to designate a basis set of Slater type orbitals, where there is one Slater orbital for each atomic function. Today's "minimal basis set," if we attempt to use the term for a gaussian basis set expansion, should mean a basis set of gaussian contracted functions that yields an energy (for the separated atom) not inferior to the one obtained for the minimal Slater basis set (69). If the gaussian expansion is so truncated as to yield an atomic energy higher than the Slater minimal basis set, the proper term for such basis is "subminimal." It is unfortunate the general misuse of the term "minimal basis sets" in current chemical literature; very often subminimal basis set are presented as minimal basis sets. For atoms containing 1s, 2s, and 2p orbitals, a minimal basis set of gaussian type corresponds to five s-type gaussian functions contracted to one 1s orbital; two s-type gaussian orbitals for the 2s orbitals and three to four p-type gaussian orbitals for the 2p orbitals, contracted (92) into one s-type functions and two p-type function, respectively. In standard notation we write that the (7/3) primitive basis set is contracted to (5,2/3) yielding a (1,1/1) contracted set. For atoms with electronic configurations $1s^2\ 2s^2\ 3s^m\ 2p^6\ 3p^n$ a minimal basis set is a (9/6) basis set contracted to (5,2,2/4,4) yielding a (1,1,1/1,1) contracted set. Decent minimal basis sets for Li to Ne are available from reference (93) and for Ne to Ar from reference (94).

Because gaussian functions at large distances from the nucleus decay to zero faster than Slater type functions, subminimal basis sets yield often erratic values in the computation of intermolecular interactions. In Table 24 we report few examples taken from literature; the notation HF

Table 24. Comparison of Intermolecular interactions with different basis sets.

System	Method	Year	Ref.	ΔE(comp) Kcal/mol	ΔE(exp) Kcal/mol
Li^+-H_2O	HF(e)	73	66	32.2	34.0
"	HF(sm)	76	a	80.0	
"	HF(m)	76	1	40.0	
Na^+-H_2O	HF(e)	73	67	24.0	24.0
"	HF(sm)	76	a	42.6	
"	HF(m)	76	a	28.9	
K^+-H_2O	HF(e)	73	67	16.6	17.0
"	HF(m)	76	a	20.3	
$Mg^{++}-H_2O$	HF(e)	78	3	75.3	–
"	HF(m)	76	a	80.0	
"	HF(sm)	76	a	117.1	
$Ca^{++}-H_2O$	HF(e)	78	3	49.0	–
"	HF(m)	76	a	47.2	
$Zn^{++}-H_2O$	HF(e	78	56	82.0	–
"	HF(m)	78	e	75.0	
"	HF(sm)	78	b	112.9	
$Zn^{++}-CO_2$	HF(m)	78	56	46.9	–
"	HF(sm)	78	b	79.4	
$Zn^{++}-Im$	HF(m)	78	56	91.0	–
"	HF(sm)	78	b	169.0	
$Li^+-O(CH_3)_2$	HF(m)	79	63	31.0	38.0
"	HF(sm)	75	c	78.0	
$Na^+-O(CH_3)_2$	HF(m)	79	62	21.6	–
"	HF(sm)	77	d	38.9	
$Li^+-O(C_2H_5)_2$	HF(m)	79	63	32.3	–
"	HF(sm)	77	d	43.2	

a. A. Pullman, H. Berthod, N. Gresh; Int.J.Quant.Chem., 10, 59 (1976)
b. D. Demoulin, A. Pullman, H. Sakar; Theo.Chim.Acta., 49, 16a (1978)
c. A. Pullman, C. Giessner-Pretter, Yu. Kruglyak, Chem.Phys. Letters, 35, 156 (1975)
d. J. F. Hinton, A. Beeler, D. Haspool, R. W. Briggs, A. Pullman, Chem. Phys. Letters, 47, 411 (1977)

or CI designate the method used in the computation, the notations (e), (m) and (sm) refer to extended, minimal and subminimal basis sets, respectively. We feel it should be unnecessary to stress that subminimal basis sets should be avoided (Table 24 might be useful to those not in agreement with the above statement). It is noted, however, that subminimal basis set, with exponents obtained by fitting experimental data, can be successfully used for geometry's optimization, if, and only if, the molecule examined does not present electronic peculiarities not present in the set of molecules used as model compounds in the orbital exponents fitting.

There are many ways to create an energetically poor basis set, and ample effort has been devoted to explore nearly all the possible avenues. We mention two ways, in addition to the one represented by the use of subminimal basis set. One way is to use (for intermolecular forces) a minimal basis set, <u>nearly uncontracted</u> a second way is to use a minimal basis set and add to it <u>polarization functions</u>. Both techniques are nearly unique in ensuring a large basis set super position error. It is noted that the literature on small molecules in the early 1960s, had repeatedly pointed out these and other similar errors.

From Table 24 we can see computations in error up to and above 100% relative to the corresponding values obtained with more realiable basis sets.

It has become nearly customary in quantum chemical computation to analyze the computed interaction energies results by some energy partitioning where the coulomb part, the exchange part, the dispersion part, etc. are carved out from the total energy; we wonder about the usefulness of partitioning energies that are up to 100% in error and even more we wonder about the physical reliability of such partitions even at a qualitative level, especially not proven that the error can be partitioned in a proportional way into the components of the total interaction.

The basis set superposition error can be <u>nearly</u> eliminated rather easily by making use of the counterpoise method presented several years ago by Boys and Bernardi (95), and rediscovered, a few times, lately for example, by W. Kolos (96). To offset most of the basis set superposition error in the computation of intermolecular forces one performs for the interaction of two molecules A and B the following set of computations: a) evaluation of E'(A) with the basis set of A and of B, centering the functions at the position selected for the A-B complex, b) evaluation of E'(B) with the basis set of B and of A, centering the functions at the positions selected for the A-B complex, and c) evaluation of E(AB) with the basis set of A and B, as standardly done.

The quantities E'(A) and E'(B) are different from those computed with the basis set either of A or of B. The interaction energy is obtained as:

$$I(A,B) = E(A,B) - E'(A) - E'(B) \qquad 4.5$$

rather than using the difference E(A,B) - E(A) - E(B). Clearly E'(A) and E'(B) are functions of the distance and orientation of A relative to B. The computational effort to compute the superposition error correction clearly increases, if we consider a complex with more than two molecules, like A-B-C. In a version of IBMOL, the problem has been nearly eliminated by simply recalling that the two electron integrals needed to compute the complex A-B (or A-B-C) at a given orientation and distance of A relative to B are exactly those needed to compute E'(A) and E'(B); therefore, a redefinition of data sets for the one and two electron intergrals is the only change one needs to add to a previous molecular program (3). In Table 25 we show an example of basis set

Table 25. Zn++-Imidazole interaction with and without counter-poise method.

R(a.u.) Zn-N	E (uncorrected) kcal/mole	E (corrected) kcal/mole
2.7	-24.74	-3.46
3.1	-79.22	-61.55
3.4	-96.56	-86.81
3.7	-96.93	-91.58
4.0	-89.58	-86.70
4.3	-80.43	-78.75
4.5	-73.41	-72.27

correction using a <u>minimal</u> basis set for the system $Zn^{++}-C_3N_2H$ (the Zn^{++} is on the nitrogen <u>lone pair</u> direction); one can notice very serious errors at short distance (R-2.7 and 3.1) an appreciable error at the equilibrium distance (R-3.7). Clearly, the correction is more dramatic for sub-minimal basis sets (see Table 24).

4.6 THE DISPERSION ENERGY

Configuration interactions or perturbation calculations of the dispersion energy are feasible only for small system since large basis sets are required. Ab-initio dispersion energies calculated using small basis sets are known to be unreliable (97). For large separations of the interacting molecules the dispersion energy can be obtained from the well known London formula (98) which relates it to the polarizabilities of the molecules. In the case of large molecules their dimensions, however, are not small in comparison with intermolecular separations of chemical interest and therefore the London formula is not applicable. For such systems the total dispersion energy can be assumed (see reference (99) and references therein) to be given by a sum of contributions due to the dispersion interaction between the bonds i and j of the interacting molecules A and B respectively.

$$E_{disp} = \sum_{i \in A} \sum_{j \in B} E_{disp}^{ij} \qquad 4.6$$

The individual contributions can be conveniently calculated using a formula derived by Claverie (99) which is essentially the London formula for the dispersion interaction of two linear molecules with anisotropic polarizabilities

$$E_{disp}^{ij} = -\frac{1}{4} \frac{U_A U_B}{U_A + U_B} \frac{1}{R_{ij}^6} Tr[\underline{T}_{ij} \underline{A}_i \underline{T}_{ij} \underline{A}_j] \qquad 4.7$$

where U_A and U_B are some average excitation energies usually related to the ionization potentials of A and B respectively, $R_{ij} = |\underline{R}_{ij}|$ where $\underline{R}_{ij}$ denotes the vector joining the midpoints of bond j in molecule B and bond i in molecule A, $\underline{A}_i$ and $\underline{A}_j$ are the polarizability tensors of the two bonds, and $\underline{T}_{ij}$ denotes the tensor

$$T_{ij} = 3(\underline{r}_{ij} \otimes \underline{r}_{ij}) - 1 \qquad 4.8$$

where $\underline{r}_{ij} = \underline{R}_{ij}/R_{ij}$.

Evaluation of the expression in the square brackets in Eq. (4.7) gives

$$TR\,[\underline{T}_{ij}\,\underline{A}_i\,\underline{T}_{ij}\,\underline{A}_j] = 6\alpha_i^T\alpha_j^T + \delta_i\alpha_j^T\,[3(\underline{r}_{ij}\cdot\underline{e}_i)^2 + 1] + \qquad 4.9$$

$$+ \delta_j\alpha_i^T\,[3(\underline{r}_{ij}\cdot\underline{e}_j)^2 + 1] + \delta_i\delta_j\,[3(\underline{r}_{ij}\cdot\underline{e}_i)(\underline{r}_{ij}\cdot\underline{e}_j) - (\underline{e}_i\cdot\underline{e}_j)]^2$$

where α^T denotes the transverse polarizability, δ the anisotropy of the polarizability defined as $\delta = \alpha^L - \alpha^T$, where α^L is the longitudinal polarizability. The indices i and j refer to the bonds in molecule A and B, respectively, and $\underline{e}_i$ and $\underline{e}_j$ are the unit vectors in the directions of the bonds indicated by their indices.

In principle both the theoretical and experimental bond polarizabilities can be used in Eq. (4.9). At present, however, only the experimental values are available (100,101). In our work standard values of α^T and δ have been employed (100,102). For U_C (C = A,B) London recommended using the ionization potentials. It has been noticed many times, however, that more realistic dispersion energies are obtained if U has a value about twice the ionization potential. For some molecules, such as H_2O or CH_4, the value of U can be determined from the experimental value of C_6, that is, of the coefficient at R^{-6} in the London formula. Using C_6 = 45.5 a.u. (103) and 150 a.u. (104), δ = 9.63 a.u. (102) and 17.54 a.u. (105) for H_2O and CH_4, respectively, one gets from the London formula $U(H_2O) = 0.653$ a.u. and $U(CH_4) = 0.650$ a.u., that is, almost identical values. In view of this the value U = 0.65 a.u. can by used; by simple scaling, the dispersion energies corresponding to other values of U can be obtained.

In the following we shall consider the dispersion energy for the cases of an atom interacting with a molecule, Ar.HCl and of two interacting molecules, $CH_4 \cdot CH_4$. Details of these computations can be found in references (106) and (107), respectively.

Let us consider first the sytem Ar.HCl; the total dispersion interaction is split into interactions between individual bonds. The interaction center of each bond is assumed to coincide with the mid-point of the bond. For our case the dispersion correction is given by the relation

$$E_{disp} = \sum_B \frac{U_{Ar} U_B}{4\,r^6\,(U_{Ar}+U_B)} \left\{ 6\alpha_{Ar}\alpha_B^T + \delta_B\alpha_{Ar}[3(\underline{r}\cdot\underline{e}_B)^2 + 1]\right\} \qquad 4.10$$

where B labels the bonds in the molecule interacting with Ar, α_x and α_x^T denotes the average and the transverse polarizability, respectively, of HCl or Ar, δ_x is the anisotropy of the polarizability, $\underline{r}$ is the radius vector from the Ar nucleus to the midpoint of bond B, $\underline{e}_B$ is the unit vector in the direction of bond B, and U_x denotes the average excitation energy of HCl or Ar.

The dispersion energy can be expanded as

$$E_{disp}(R,\theta) = \sum_{n=0} E_{dn}(R) P_n(\cos\theta) \qquad 4.11$$

where R denotes the Ar...Cl distance and θ is the ArClH angle. Expansion (4.11) is usually limited to the first three terms and E_o is called the isotropic potential.

Restricting the expansion (4.11) to the first three terms, and using the polarization approximation and the multipole expansion one obtains (108,109)

$$E_{disp} = -C_6 R^{-6}(1+a_1 R^{-1} P_1(\cos\theta)+a_2 P_2(\cos\theta)) \quad\quad 4.12$$

where C_6 is the familiar constant which denotes the isotropic dispersion interaction

$$C_6 = -(3/2)\alpha_A \alpha_B\, U_A U_B (U_A + U_B)^{-1} \quad\quad 4.13$$

The third term in Eq. (4.12), proportional to P_2 is included in our Eq. (4.10). Since, however, the anisotropy of the polarizability of HCl is small, this term introduces only a very weak dependence of E_{disp} on the angle θ. The term, proportional to P_1, seems to be more important. Explicitly there is no such term in Eq. (4.10). However, it is contained in this equation implicitly, if r is treated not as the Ar...Cl distance (R) but as a distance from the Ar nucleus to the HCl midpoint. If the HCl bond length, r_{HCl}, is small in comparison with r, we have

$$r^{-6} = \sim R^{-6}(1+3r_{HCl}R^{-1}\cos\theta) \quad\quad 4.14$$

Hence implicitly Eq. (4.10) contains a term proportional to P_1 and the term has the correct power of R, that is, R^{-7} (108,109). This partly justifies our approach in which even the hydrides we identify the interaction center with the bond midpoint.

In principle the value of a_1 in Eq. (4.12) can be determined from the properties of the interacting subsystems. It has been found (110), however, that better results are obtained if one introduces a center of the dispersion interaction which does not coincide with the center mass of HCl. The location of the interaction center is then determined either from theoretical considerations or by fitting the results to some experimental data. This is analogous to our approach where the main difference consists of the fact that choosing r rather than R as the separation of the interacting subsystems, we fix the value of a_1 in Eq. (4.12). Obviously, we can also determine the position of the interaction center by fitting the results to experimental data. For this purpose the data related to large Ar...Cl separations would be desirable which, however, are not available. For this reason, and to avoid new parameters, we localized the interaction center in the bond midpoint, keeping in mind that the angular dependence of E_{disp} may be exaggerated.

Having rationalized the use of r rather than R in Eq. (4.10) let us now discuss the parameters appearing in this equation.

To determine U_{Ar} we use C_6 and α_{Ar} obtained by Tang et al (111) and we obtain $U_{Ar} = 0.727$ a.u. For HCl the values of U_x have been assumed to be proportional to the ionization potentials. Using U_{H_2O} as the reference value this gives $U_{HCl} = 0.66$ a.u. The constants C_6 for Ar.HCl can be now easily determined. With $\alpha_{HCl} = 16.85$, $\delta_{HCl} = 2.10$ a.u. (112) one gets $C_6 = 101$ a.u.

From studies of the dispersion interaction, in particular between rare gas atoms, it is well known that the first term of the multipole expansion proportional to R^6 is not sufficient to yield the interaction energy in agreement with experiment. On the other hand, since this is an asymptotic expansion, if higher terms are included the problem of divergence arises, especially for distances smaller than that for the Van der Waals minimum. To avoid this difficulty simple damping procedures have been proposed (113) for the dispersion energy restricted to the first three terms of the multipole expansion. To get at least very rough estimates of the importance of the higher terms of Ar.HCl we have followed the same procedure. The dispersion energy was calculated as

$$E_{disp}(10) = [E_{disp}(6) + C_8 r^{-8} + C_{10} R^{-10}] f(r) \qquad 4.15$$

where $E_{disp}(6) = E_{disp}$ is defined in Eq. (4.11) and $f(r)$ is the damping factor used by Duguette et. al, (113) in the form recommended by Ahlrichs (114).

The values of C_8 and C_{10} for Ar.HCl have been obtained by interpolation from those for Ar.Kr and Ar.Xe (111) assuming their changes to be proportional to those of C_6. The resulting values are $C_8 = 2370$ and $C_{10} = 6790$ a.u. Since the contributions to the dispersion energy due to the terms proportional to C_8 and C_{10} are not very large, uncertainties in their values do not affect significantly the final interaction energies. For the same reason we think it is justified to assume, as done in Eq. (4.15) the angular dependence of the higher terms to be due to the dependence of r on θ. A more reliable determination of the angular dependence of C_8, although in principle possible, also fails to give accurate results (115).

The SCF binding energies computed for the complex Ar.HCl at various Ar ...Cl distances, R, and Ar.ClH angles, θ, are listed in Table 26. We

Table 26. Energies of Ar.HCl for Various Geometrical Configurations of the System*

ϑ	R	$\Delta E'$	ΔE	ϑ	R	$\Delta E'$	ΔE
0	6.50	4.487	3.521	90	7.75	0.077	0.053
0	7.00	1.631	0.993	90	8.00	0.042	0.028
0	7.50	0.350	0.191	90	9.00	0.003	0.002
0	7.75	0.588	0.048	120	6.50	1.028	0.839
0	8.00	0.206	-0.018	120	7.00	0.359	0.272
0	9.00	0.021	-0.034	120	7.50	0.117	0.082
30	6.50	2.883	2.212	120	7.75	0.065	0.043
30	7.00	1.102	0.690	120	8.00	0.035	0.023
30	7.50	0.412	0.179	120	9.00	0.002	0.001
30	7.75	0.249	0.079	150	6.50	0.740	0.581
30	8.00	0.148	0.028	150	7.00	0.252	0.181
30	9.00	0.016	-0.011	150	7.50	0.079	0.053
60	6.50	1.448	1.135	150	8.00	0.023	0.017
60	7.00	0.540	0.388	150	9.00	0.001	0.004
60	7.50	0.193	0.124	180	6.50	0.581	0.431
60	7.75	0.113	0.068	180	7.00	0.193	0.125
60	8.00	0.065	0.036	180	7.50	0.059	0.032
60	9.00	0.006	0.001	180	7.75	0.032	0.015
90	6.50	1.169	0.956	180	8.00	0.017	0.007
90	7.00	0.416	0.318	180	9.00	0.001	0.000
90	7.50	0.138	0.098				

*R in atomic units, interaction energies, $\Delta E'$ and ΔE in Kcal/mole.

list two interaction energies $\Delta E'_{SCF}$ and ΔE_{SCF} in kcal/mole; the former was calculated in the traditional way, that is, by using the energies of infinitely separated Ar and HCl ($E_{Ar} = -525.864988$ and $E_{HCl} = -459.234804$ a.u.). To obtain ΔE_{SCF} the CP method (95) was employed, that is the energies E_{Ar} and E_{HCl} listed for each configuration in Table 26 were subtracted from the total energy of the system for this configuration.

In Table 27 the dispersion energies are listed. For each configuration we give two values: $E_{disp}(6)$ calculated from Eq. (4.10) and $E_{disp}(10)$ obtained using Eq. (4.15).

Total interaction energies for $\theta = 0°$ calculated using ΔE_{SCF} in Eq. (4.10) are shown in Figure 12 where they are compared with three other curves obtained by fitting analytical expressions to some experimental data. The analytical expressions were of the form

$$V(R,\theta) = V_0(R) + a_1 V_1(R) \rho R^{-1} P_1(\cos\theta) + a_2 V_2(R) P_2(\cos\theta) \qquad 4.16$$

with V_1 represented by modified Buckingham potentials

Table 27. Dispersion Energies for Various Geometrical Configurations of Ar.HCl*.

ϑ	R	E(6)	E(10)	ϑ	R	E(6)	E(10)
0	6.5	-2.991	-2.373	90	8.0	-0.222	-0.295
0	7.0	-1.741	-1.696	90	9.0	-0.111	-0.147
0	7.5	-1.059	-1.175	90	10.0	-0.060	-0.077
0	8.0	-0.670	-0.805	120	6.5	-0.465	-0.590
0	9.0	-0.294	-0.380	120	7.0	-0.312	-0.406
0	10.0	-0.142	-0.186	120	7.5	-0.214	-0.283
30	6.5	-2.360	-2.077	120	8.0	-0.150	-0.199
30	7.0	-1.404	-1.463	120	9.0	-0.078	-0.102
30	7.5	-0.870	-1.008	120	10.0	-0.043	-0.055
30	8.0	-0.558	-0.691	150	6.5	-0.348	-0.447
30	9.0	-0.251	-0.328	150	7.0	-0.237	-0.310
30	10.0	-0.124	-0.162	150	7.5	-0.165	-0.217
60	6.5	-1.356	-1.440	150	8.0	-0.118	-0.154
60	7.0	-0.844	-0.994	150	9.0	-0.063	-0.080
60	7.5	-0.544	-0.682	150	10.0	-0.036	-0.044
60	8.0	-0.361	-0.469	180	6.5	-0.315	-0.407
60	9.0	-0.171	-0.227	180	7.0	-0.216	-0.282
60	10.0	-0.088	-0.115	180	7.5	-0.152	-0.198
90	6.5	-0.746	-0.898	180	8.0	-0.180	-0.141
90	7.0	-0.485	-0.617	180	9.0	-0.058	-0.074
90	7.5	-0.324	-0.425	180	10.0	-0.033	-0.041

*R in atomic units, E in Kcal/mole

$$V_i(R) = \frac{E_i}{\alpha_i - 6}\left\{ 6 \exp[\alpha_i(1-\frac{\rho_i}{R})] - \alpha_i(\frac{\rho_i}{R})^6 \right\} \qquad 4.17$$

Neilsen and Gordon (109) assumed V_i to be independent of i, and used for E, α and ρ the values determined (116) in studies of transport data. The parameters a_1 and a_2 were calculated by a least squares fit to the experimental infrared line widths, line shifts and proton magnetic relaxation times. This potential is labelled NG in Figure 12.

Recently Holmgren et al (117) have used experimental data of molecular beam resonance spectroscopy (118,119) to determine parameters in Eq. (4.17). They assumed $E_1 = E_2$, $\alpha_1 = \alpha_2$ and $\rho_1 = \rho_2$. In addition they found α_o and E_o to be highly correlated. For this reason they determined two potentials. In the first they did not vary the value of α_o

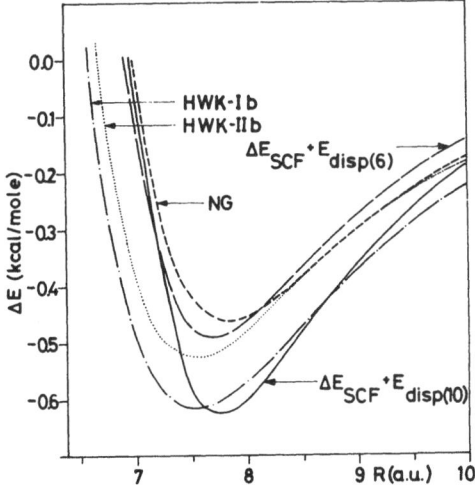

Figure 12. Ar.HCl interaction energies for θ = 0°; the computed curves are ΔE(SCF)+E(dis.,6) and ΔE(SCF)+E(dis.,10); Neilsen and Gordon data are labelled NG and two of the potentials by Holmgreen, et al are labeled HWK-IB and HWK-IIb (see text).

but assumed $\alpha_0 = 13.5$ used also by Neilsen and Gordon (109). This potential is labelled HWK-Ib in Figure 12. In the second potential they fixed E_0 at the value $E_0 = 133$ cm^{-1} determined in elastic scattering experiments (120) and varied the remaining parameters. This potential is labelled HWK-IIb in Figure 12.

Figure 12 shows that the Ar.HCl potential is fairly sensitive to the experimental data used to determine adjustable parameters in the potential and also to additional assumptions with regard to the parameters (difference between HWK-Ib and HWK-IIb). The theoretical $\Delta E_{SCF} + E_{disp}(6)$ potential agrees fairly well with the HWK-IIb and NG potentials, whereas $\Delta E_{SCF} + E_{disp}(10)$ is more similar to HWK-Ib. Since, however, HWK-IIb better reproduces the experimental data our $\Delta E_{SCF} + E_{disp}(6)$ potential seems to be more reliable than $\Delta E_{SCF} + E_{disp}(10)$.

There are several possible reasons of the above conclusion. Our values of C_8 and C_{10} or, even more, the form of the damping factor may be subjected to criticism. It is, however, unlikely that the contributions due to C_8 and C_{10} are negligible. Therefore the potential $\Delta E_{SCF} + E_{disp}(6)$ may owe its satisfactory form either to an overestimation of $E_{disp}(6)$ caused by our choice of the interaction center in the midpoint of the HCl bond, or by an underestimation of ΔE_{SCF} due to the incompleteness of the basis sets that might be underestimated by as much as 0.05 kcal/mole.

Equivalent studies have been performed for the Ar.HF and the Ar.H$_2$O systems (107).

These studies show that the SCF approach with judiciously chosen minimal basis sets and the counterpoise method, combined with a simple calculation of the dispersion energy, can be successfully employed to study van der Waals molecules.

Let us now consider the example of CH$_4$ interacting with CH$_4$ (106). One extended and two minimal basis sets have been used. The extended basis set (11,7,1/6,1) contracted to (4,3,1/3,1), was the same as that used by Tosi, et al (121) in their study of the CH$_4$.H$_2$O interaction. Both minimal basis sets were of the (5,2/3) type, contracted obviously to (1,1/1). One of them, denoted as "old," was taken from van Duijneveld's work (122). The second, denoted as "new," differed from the first one in two respects (123): the contraction coefficients in the hydrogen 1s orbital have been optimized for CH$_4$, and the exponents in the carbon valence orbitals (2s and 2p) have been multiplied by a scale factor equal to 1.05. For the experimental geometry of the methane molecule ($R_{CH} = 2.0665$ a.u. and tetrahedral bond angle) the energies resulting from those basis sets are E = -40.210449, -39.98592 and -40.10259 a.u., respectively. The best SCF energy of CH$_4$ obtained using a very extended basis set is E = -40.2136 a.u. (124). In addition, less extensive computations have been made using the STO-4G basis set (125) which for an isolated CH$_4$ molecule gives E = -40.00710 a.u.

Six mutual orientations of the methane molecules have been considered which will be labelled A,...,F. They are shown in Figure 13. For each orientation the energy was calculated for several C...C distances.

Let us first point out that extended basis set is still not flexible enough to give a negligible basis set superposition error (BSSE). At R = 7.5 the error amounts to 0.25 and 0.21 kcal/mole for configuration A and D, respectively. For configuration A and $R_{CC} = 6$ it amounts to 0.54 kcal/mole. It is also seen that for configuration A the BSSE is sufficiently large to give a spurious potential minimum which disappears

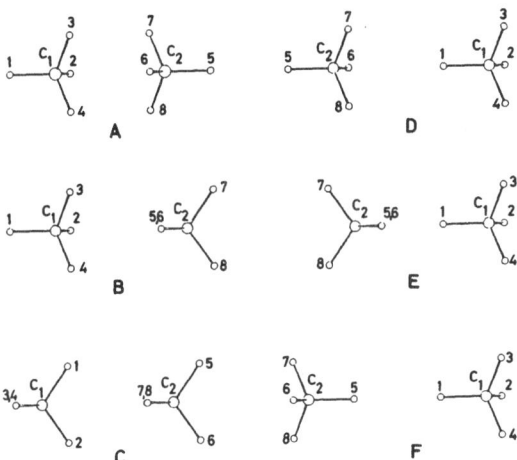

Figure 13. Six mutual orientations of the methane molecules considered in the present work.

when the CP method is employed. Even for configuration D, in which the octupole-octupole interaction is attractive (see below), the extended basis set calculations, when carried out using the CP method, give no potential minimum for R≤9 a.u. The negative interaction energy obtained in the traditional approach for R = 9 is seen to be a basis set effect. On this basis we assume that also the minima in configurations B and E are spurious. In the case of the minimal basis sets the interaction energies for R = 9 in configuration D remain somewhat attractive also with the CP method. This may be due to larger polarities of the CH bonds (123), and hence stronger octupole-octupole attraction resulting from these basis sets.

The minimal basis set denoted as "old" gives a large BSSE, and the interaction energies calculated in the traditional way using this basis set are seen to be unreliable. If the BSSE is removed using the CP method one gets significantly improved interaction energies which, however, are still considerably smaller (less repulsive) than those resulting from the extended basis set.

On the other hand the interaction energies calculated using the new minimal basis set are remarkably good. In this case the BSSE is small and the interaction energies E for small and intermediate distances are close to those obtained with the extended basis set. For large intermolecular distances the results of the minimal basis set calculations are likely even more reliable since they give no spurious minima for the repulsive configurations.

Let us now briefly comment on the interaction energies calculated using the STO-4G basis set (106). For configuration A they are slightly less repulsive than the energies obtained with the new minimal basis set, that is, more similar to those resulting from the extended basis set, but in contrast to the latter the STO-4G basis set gives no spurious minimum and a small BSSE. At R_{CC} = 7.5 the BSSE amounts to 0.03 kcal/mole. For configuration F and large distances the energies obtained with the STO-4G basis set are practically indistinguishable from those calculated using the new minimal basis set; for smaller distances (R_{CC}= 7) they are slightly more repulsive. In general the STO-4G basis set

gives interaction energies which are very close to those resulting from
the new minimal basis set. A typical calculation with the STO-4G basis
takes, however, about twice as much of computer time as a calculation
with the (5,2/3) minimal basis set. In a previous SCF calculation of
the $CH_4 \cdot CH_4$ interaction energy by Catlow et al. (126) configurations A,
D and F were considered. The calculations however, were performed
mostly for shorter C...C distances than in the present work. For configuration A and F the interaction energies calculated by Catlow, et
al. are somewhat more and less repulsive, respectively, than our results
obtained with the extended basis set. They also report a potential minimum for configuration A at $R_{CC} \sim 8.5$ a.u. which indicates that their
energies are affected by a nonnegligible basis set superposition error.
The potential energy curves calculated for all configurations using the
extended basis set are shown in Figure 14. Those resulting from the new
minimal basis set are practically identical. As examples we show in
Figure 14 the potential energy curves for configurations A, D and F,
respectively. The solid lines represent the interaction energies calculated in the traditional way, while the broken lines denote E_{int}^{CP},

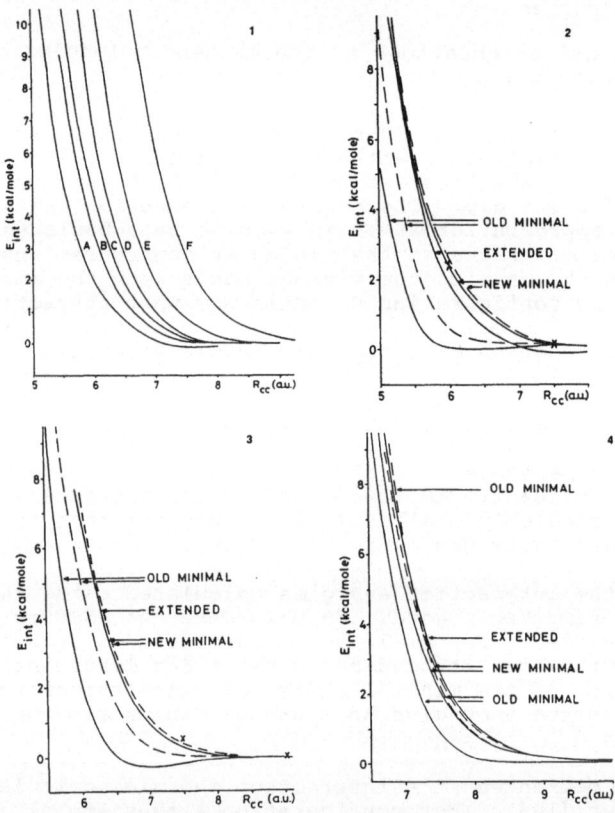

Figure 14. Top Left Insert. $CH_4 \cdot CH_4$ interaction energies calculated
using the extended basis set for six orientations shown in Fig. 13. Top
Right Insert. Interaction energies for configuration A calculated using
various basis sets. The two crosses at R=6 and 7.5 a.u. show the energies calculated using the extended basis set and the CP method. Bottom
Left Insert. Interaction energies for configuration D calculated using
various vasis sets. Bottom Right Insert. Interaction energies for
configuration F calculated using various basis sets. The broken lines
show results obtained using the CP method.

that is, the interaction energies calculated using the counterpoise method. The four energy values for A and D (two for each) which have been obtained using the extended basis set and the CP method are marked by crosses. The figures show clearly a very close similarity of the curves calculated using the extended basis set and the new minimal basis set. For configuration D, the E_{int} curves obtained with the two basis sets are undistinguishable. In the case of the new minimal basis set one also notices a close similarity between E_{int} and E_{int}^{CP} curves, that is, a small BSSE.

In Tables 28 to 30 we give the atomic populations and the energies of the molecular orbital valency states (MOVS)(54,63) calculated with the extended basis set. For each configuration of the methane molecules

Table 28. Atomic Populations for Several Configurations of $CH_4 \cdot CH_4$

Conf.	CC	qC1	qH1	qH2	qH3	qH4	qC2	qH5	qH6	qH7	qH8
B	5.5	6.268	0.972	0.928	0.916	0.916	6.239	0.936	0.935	0.945	0.945
B	8.5	6.299	0.932	0.923	0.923	0.923	6.293	0.924	0.925	0.929	0.929
D	6.0	6.247	0.938	0.939	0.939	0.939	6.324	0.958	0.906	0.906	0.906
D	8.0	6.281	0.934	0.929	0.929	0.929	6.303	0.939	0.919	0.919	0.919
E	6.0	6.249	0.955	0.940	0.938	0.938	6.258	0.928	0.929	0.933	0.933
E	8.0	6.285	0.934	0.928	0.928	0.928	6.290	0.925	0.925	0.929	0.929

Table 29. Energies of Molecular Orbital Valency States (MOVS) for Several Configurations of $CH_4 \cdot CH_4$

Conf.	CC	C1	H1	H2	H3	H4	C2	H5	H6	H7	H8
B	5.5	−36.2168	−0.2379	−0.2329	−0.2279	−0.2279	−36.1901	−0.2341	−0.2358	−0.2316	−0.2316
B	8.5	−36.2101	−0.2284	−0.2259	−0.2256	−0.2256	−36.2040	−0.2257	−0.2258	−0.2273	−0.2273
D	6.0	−36.1711	−0.2422	−0.2294	−0.2311	−0.2311	−36.2575	−0.2323	−0.2247	−0.2260	−0.2260
D	8.0	−36.1881	−0.2285	−0.2267	−0.2267	−0.2267	−36.2163	−0.2297	−0.2251	−0.2251	−0.2251
E	6.0	−36.1866	−0.2457	−0.2298	−0.2295	−0.2295	−36.2194	−0.2331	−0.2328	−0.2283	−0.2283
E	8.0	−36.1925	−0.2289	−0.2265	−0.2265	−0.2265	−36.2053	−0.2265	−0.2265	−0.2272	−0.2272

Table 30. Atomic Populations and Energies of MOVS For Several Configurations of $CH_4 \cdot CH_4$

Conf.	R(CC)	q				MOVS	
		C1	H1	H3	C1	H1	H3
A	5.0	6.248	0.987	0.922	−36.2416	−0.2423	−0.2314
A	8.0	6.293	0.940	0.022	−36.2067	−0.2308	−0.2256
C	5.5	6.282	0.920	0.939	−36.2243	−0.2331	−0.2301
C	8.0	6.294	0.924	0.929	−36.2056	−0.2273	−0.2256
F	6.5	6.250	0.956	0.931	−36.2074	−0.2456	−0.2272
F	9.0	6.289	0.935	0.925	−36.1931	−0.2303	−0.2257

only two sets of results are given: one corresponding to a relatively short C...C distance, and one for a large distance for which the intermolecular interaction is already very weak. The numbering of atoms is given in Figure 13.

It is seen that due to the interaction the outer hydrogen atoms of the dimer are always more negative than in an isolated methane molecule, and the carbon atoms are more positive, except the carbon atom in the proton acceptor molecule is configuration D. In the latter case the inner three hydrogen atoms become considerably more positive than in an isolated CH_4 molecule and the polarity of the CH bond increases. For larger intermolecular distances, however, also in this orientation the carbon atom becomes more positive.

The only noticeable charge transfer occurs to the proton donor molecule in configuration E. For shorter distances (R_{CC} = 5.5) an analogous charge transfer occurs in configuration D.

An interesting relationship can be noticed between the net charges q_x, or the changes Δq_x due to the interaction and the energies of MOVS. This is most clearly seen in the case of the hydrogen atoms. If the hydrogen atom under consideration is relatively far away from the second methane molecule, that is, if it is an outer hydrogen atom in the dimer, or is it an inner hydrogen atom but the intermolecular distance is large, the energy of its MOVS increases approximately linearly with increasing value of Δq_H. Thus if a given environment decreases the energy of the valency state of a hydrogen atom, it increases the probability of finding an electron in the vicinity of this atom. On the other hand if the hydrogen atom under consideration is relatively close to the second molecule, that is, if it is an inner hydrogen atom in the dimer and the intermolecular distance is small, the energy of its MOVS also increases with increasing Δq_H, however, for a given Δq_H it has a lower value than in the former case.

We can now discuss the dispersion correction. Using a London-type formula and experimental values of bond polarizabilities the energy of the dispersion interaction can be calculated approximately as a sum of contributions due to bond-bond interactions. Following (127) we have used

Table 31. Dispersion Energies for Various Orientations of $CH_4 \cdot CH_4$ (in Kcal/mole).

Conf. R	A	B	C	D	E	F
5.0	-6.89	-7.77	-8.56	-9.99	-11.69	-18.29
5.5	-3.82	-4.22	-4.59	-5.17	-5.86	-8.16
6.0	-2.23	-2.43	-2.61	-2.87	-3.18	-4.10
6.5	-1.37	-1.47	-1.56	-1.69	-1.84	-2.25
7.0	-0.87	-0.92	-0.98	-1.04	-1.12	-1.32
7.5	-0.57	-0.60	-0.63	-0.66	-0.71	-0.81
8.0	-0.38	-0.40	-0.42	-0.44	-0.47	-0.52
8.5	-0.26	-0.28	-0.29	-0.30	-0.31	-0.35
9.0	-0.19	-0.19	-0.20	-0.21	-0.22	-0.24
9.5	-0.13	-0.14	-0.14	-0.15	-0.15	-0.17
10.0	-0.10	-0.10	-0.11	-0.11	-0.11	-0.12

for the average excitation energy the value $U = 0.66$ a.u., and standard value (100) of the transverse bond polarizability $\alpha_{CH}^T = 3.91$ a.u. and of its anisotropy $\delta_{CH} = 1.42$ a.u. The resulting dispersion energies are listed in Table 31. With increasing intermolecular separation the anisotropy of the dispersion interaction becomes progressively less important. For $R_{CC} \geq 9$ the results of Table 31 are already very close to the isotropic dispersion energy calculated as the first term in the expansion

$$E_{disp} = \sum_{n=1}^{3} C_{2(n+2)} R^{-2(n+2)} \qquad 4.18$$

With $C_6 = 150$ a.u. (128) Eq. (4.18) gives for $R_{CC} = 9$, $E_{disp} = 0.18$ kcal/mole. Since for large distances the E_{int}^{CP} values calculated with the new minimal basis set are more reliable than E_{int} obtained with the extended one, we use the former to get the total interaction energies

$$\mathcal{E}_{int} = E_{int}^{CP} + E_{disp} \qquad 4.19$$

The results obtained for configurations A, D and F using E from Table 31 are shown in Figure 15 where the average experimental curve, labelled MS, obtained from thermophysical data (129) is also presented. It is clearly seen that no averaging of the theoretical energies can produce a curve in agreement with the experimental one. As a possible reason for this failure one may regard too small values of the dispersion energy which in the London-type formula is represented only by the first term in the multipole expansion.

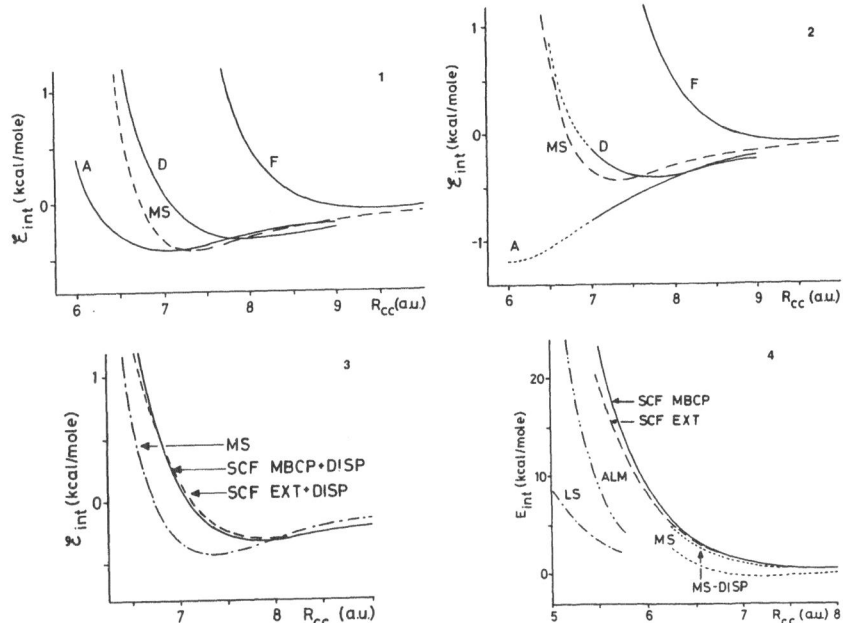

Figure 15. Top Left Insert. Interaction energies for three orientations of $CH_4 \cdot CH_4$ obtained by adding to the SCF energy the dispersion energy calculated from a London-type formula as a sum of bond-bond contributions. The broken line (MS) represents the experimental energy. Top Right Insert. Interaction energies for three orientations of $CH_4 \cdot CH_4$ obtained by adding to the SCF energy the dispersion energy calculated from Eq.(4.19). In the region where the convergence of the multiple expansion is not satisfactory the results are shown by dotted lines. The broken line (MS) represents the average experimental interaction energy. Bottom Inserts: see text.

The first term in the multipole expansion of the dispersion energy calculated as a sum of bond-bond contributions comprises obviously more than just the first term in an analogous expansion for two CH_4 molecules given by Eq. (4.18). It already includes contributions which in Eq. (4.18) are obtained with n > 1. It also includes part of the orientation dependent dispersion energy which in the lowest order is proportional to R^{-7} (130) and is absent in Eq. (4.18). However, to get some insight into the importance of various terms in the dispersion energy expansion we may calculate the total interaction energies of $CH_4 \cdot CH_4$ by using Eq. (4.18) in Eq. (4.19) with the C_m constants (m = 6,8,10) from the work by Matthews and Smith (128). In this way the orientation dependent part of the dispersion energy will be lost. One can see, however, from Table 31 that for $R_{CC} > 7$ the anisotropy of E_{disp} is relatively small, and for $R_{CC} < 7$ the results are unreliable because in this region the convergence of the multipole expansion Eq. (4.18) is not satisfactory. The $CH_4 \cdot CH_4$ interaction energies obtained by using in Eq. (4.19) the dispersion energies calculated from Eq. (4.18) are presented in Figure 29 and are seen to be much more consistent with the experimental curve than those of Figure 28. Both figures show that for each orientation of the CH_4 molecules there is a Van der Waals minimum in the energy of their interaction. The parameters of the minimum depend, however, very strongly on the orientation of the molecules. The equilibrium C...C separation may vary from 6 to 10 a.u., and the binding energy may change by an order of magnitude with the change of the orientation.

To conclude this section on the dispersion energy correction, we feel it has been proven that a reliable and easy technique has been proposed, valid for chemical systems of large size. The problem can be formulated differently by using a function of the Hartree-Fock density, as explained in the first chapter; however, the above perturbational approach can be considered a useful alternative, until the time when more reliable functionals will be available.

4.7 THREE AND MANY BODY CORRECTIONS

In the previous sections the total intermolecular interaction is approximated by pairwise additive potentials, $V(i,j)$. It is known, however, that the exact intermolecular interaction for n-body systems must be expanded in a series of two, three, ...n-body terms as given in Eq. (4.1). From preliminary quantum mechanical studies on small ion-water cluster (131) it appears that the above series converges rather slowly and the terms have alternate signs. The three-body correction has been analyzed in detail for the system lithium ion-water-fluoride ion (132) and water-water-water (133). Two examples are here analyzed in same detail in order to show how to obtain the most important part of the many body correction; the examples are $(H_2O)_3$ and $Li^+(H_2O)_2$.

Nonadditivity of interactions between closed shell atoms has been extensively studied. Little, however, is known about the nonadditivity of interactions between molecules, about its magnitude and its influence on the properties of molecular clusters. A few SCF studies have been made of the nonadditivity of interactions between water molecules. Hankins et al. (133) have calculated the nonadditive component of the interaction energy of $(H_2O)_3$ for four different geometries and several intermolecular distances. With less sophisticated basis sets the nonadditivity of interaction in water trimers has also been studied by Del Bene and Pople (134), and in water-methanol trimers by Del Bene (135). Kistenmacher, et al. (135) have used extended basis sets to determine the many-body contributions to the interaction energy of the most stable $(H_2O)_3$ and $(H_2O)_4$ clusters. The results of Hankins, et al. (133) and of Kistenmacher et al. (136) agree in indicating that for the water trimer at equilibrium geometries the non additivity, in absolute value, is of the order of 1 kcal/mole or less. The results of Del Bene and Pople (134) show the nonadditivity to be considerably larger but these results are partly due to the assumed geometries of the trimers where the O...O distances were considerably shorter than those determined with more extended basis sets (133,136). However, even if this effect is taken into account, the nonadditive energy component still remains strongly dependent on the basis set employed and the values resulting from small basis sets seem to be unreliable. On the other hand only for systems of the size of $(H_2O)_3$ it is at present possible to get reliable - hopefully - three-body energies using fairly extended basis sets. If, however, one molecule in the trimer is considerably larger, the calculation becomes feasible only if minimal basis sets are employed, and hence the reliability of the results is questionable. In addition, since the nonadditive component of the interaction energy is small and is obtained as a difference of much larger numbers, the error may be larger than the effect itself and there is no way of even predicting its sign.

In Clementi et al. work (137) twenty-nine different geometrical configurations of $(H_2O)_3$ have been studies using the SCF method. Particular attention has been paid to two configurations for which large values of the non-additive contributions to the interaction energy have been obtained. One of them is stronly attractive and represents the stable cyclic trimer (136), whereas the second configuration is repulsive. Both are shown in Figure 16. For these two configurations a more detailed study of the basis set effect on the interaction has been made (137).

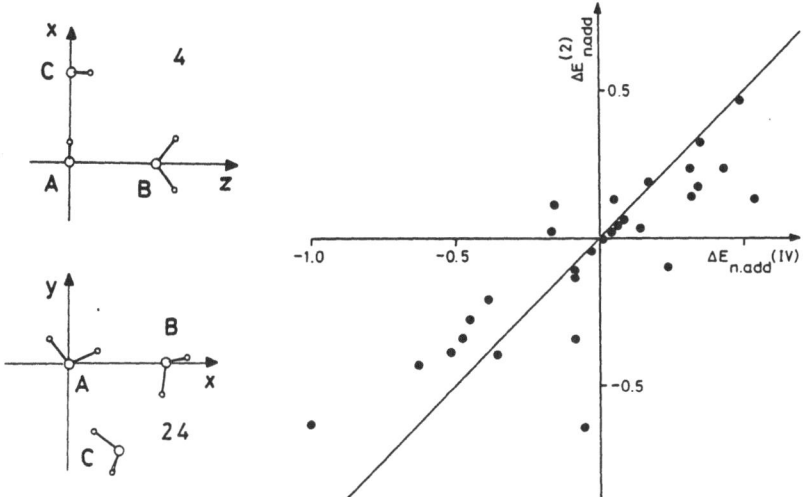

Figure 16. Configurations 4 and 24 of $(H_2O)_3$ (left inserts, top and bottom) and correlation between nonadditive components of the interaction energy for various configurations of $(H_2O)_3$ calculated using the SCF method with extended basis sets and Eqs. (4.21) to (4.23) for the induction energy (in Kcal/mole).

The total interaction energies denoted by ΔE_{ABC} for these two configurations designated in (137) as 24 and 4, as well as the two-body interaction energies, ΔE_{KL} with K,L = A,B,C and the nonadditive interaction energy defined as

$$\Delta E_{n.add} = \Delta E_{ABC} - \Delta E_{AB} - \Delta E_{AC} - \Delta E_{BC} \qquad 4.20$$

are listed in Table 32. For each basis set we give two sets of interaction energies. If the symbol of the basis set is provided with an asterisk, that is, M* with M = I,II,III and IV, the corresponding interaction energies have been calculated using the counterpoise (CP) method (95).

By inspecting the energies listed in Table 32 for configuration 24 we conclude that the basis set II (small basis plus polarization functions) gives the largest BSSE and poor results even with the CP method. This poor performance is probably due to the unrealistic charge distribution resulting from the basis set. Basis set III is considerably more flexible and hence gives a smaller BSSE. However, also in this case, the poor charge distribution yields unreliable interaction energies.

On the other hand, the basis set I, hereafter called minimal basis set, gives a realistic charge distribution in a single water molecule. Therefore, using this basis set and eliminating the BSSE by employing the CP method one gets interaction energies which are remarkably close to those obtained with the basis set IV, hereafter called extended basis set.

In the case of configuration 4 all two-body interactions are repulsive and the performance of the minimal basis set, although satisfactory, is

Table 32. Interaction Energies in $(H_2O)_3$ Calculated Using Various Basis Sets (in Kcal/mole) [1]

Config.	Basis Set[2]	ΔE	ΔE	ΔE	ΔE	ΔE
24	I	-19.83	-6.05	-5.75	-6.47	-1.56
	I*	-12.96	-3.77	-3.93	-4.11	-1.16
	II	-21.29	-6.71	-5.97	-7.06	-1.56
	II*	-11.42	-3.30	-3.48	-3.64	-1.00
	III	-20.12	-6.04	-6.15	-6.30	-1.64
	III*	-16.05	-4.72	-4.87	-4.99	-1.47
	IV	-13.57	-4.15	-4.06	-4.36	-1.00
	IV*	-12.73	-3.79	-3.75	-3.97	-1.24
4	I	3.64	3.63	-0.16	0.05	0.11
	I*	9.13	4.69	4.21	0.24	-0.02
	II	2.46	2.69	-0.62	0.08	0.32
	II*	10.27	4.42	5.34	0.48	0.03
	III	7.02	4.98	1.80	0.06	0.18
	III*	10.49	5.75	4.41	0.20	0.13
	IV	11.19	5.53	4.81	0.37	0.49
	IV*	12.41	6.04	5.55	0.44	0.38

[1] Atomic unit of energy = 627.5 Kcal/mole.

[2] Definition of basis sets I, II, III, IV in Reference 137; an asterisk denotes that the counterpoise method was employed.

somewhat less spectacular than for configuration 24. In addition, the basis set III is seen to be superior to the minimal basis set in spite of its giving poor charge distribution. This is likely to be because the electrostatic interaction energy for configuration 4 is relatively small.

For the nonadditive component of the interaction energy, in the case of configuration 24, reasonable values have been obtained with all three approximate basis sets. The CP method changes these values in the right direction giving the final nonadditivities a fairly satisfactory agreement with the value obtained using the extended basis set. In the case of configuration 4 the values of $\Delta E_{n.add}$ obtained with approximate basis sets are rather poor and the CP method changes them in the wrong direction. On the other hand it is also seen that in absolute value the differences between $\Delta E_{n.add}$ values obtained using the extended and more approximate basis sets are roughly the same for both configurations and amount to a few tenths of a kcal/mole.

With regard to the basis sets the results give no justification for using, in problems of intermolecular interactions, basis sets of intermediate quality. If for computational or economical reasons very extended basis sets cannot be used, it seems that in many cases <u>well balanced minimal basis sets</u> can give results which are even more accurate than those obtained with basis sets of intermediate quality. For a still more complete assessment of the various basis sets it may be mentioned that the CPU times needed for an average $(H_2O)_3$ calculation were in the ratio 1:25:8:250 for the I, II, III, and IV basis sets, respectively. We refer to the paper by Clementi, et al. (137) for a detailed discussion on the remaining twenty-seven configurations for $(H_2O)_3$.

One may expect that an important contribution to the nonadditive interaction energy between H_2O molecules is due to the second-order long-range induction interaction. Approximate values of this contribution can easily be obtained from the energy of the induction interaction between the induced dipoles of individual bonds of one molecule and the point charges localized on atoms of all other molecules (99). Thus for any cluster of N molecules the total induction energy can be expressed as as

$$E_{ind}^{(2)} = \sum_{K=1}^{N} E_{ind.K}^2 \qquad 4.21$$

with

$$E^{(2)}_{ind.K} = -\frac{1}{2} \sum_{i \in K} [\alpha^T_i (\xi_i \cdot \xi_j) + \delta_i (\xi_i \cdot e_i)^2] \qquad 4.22$$

where α^T_i and δ_i denote the transverse polarizability and its anistropy, respectively, of bond i in the molecule K, ξ_i is the electric field vector at the midpoint of bond i

$$\xi_i = \sum_j \frac{q_j}{R^3_{ij}} R_{ji} \qquad 4.23$$

where the summation estends over all atoms with point charges q_i, of all other molecules in the cluster different from K, R_{ji} denotes the radius vector from the atom j to the midpoint of the bond i, and e_i in Eq. (4.22) is the unit vector in the direction of bond i. From Eqs. (4.21) to (4.23) the nonadditive component of $E^{(2)}_{ind}$, which will be denoted by $\Delta E^{(2)}_{n.add}$ can easily be extracted. It is given by that part of $E^{(2)}_{ind}$. which depends on the products $q_j q_k$, with j and k denoting atoms belonging to different molecules.

Using standard values of the bond polarizabilities, α^T_{OH} = 3.91, δ_{OH} = 1.42 a.u., and the point charges q_O = 0.682, q_H = 0.341 obtained from the extended basis set calculation, $\Delta E^{(2)}_{n.add}$ has been calculated for all configurations of $(H_2O)_3$. Figure 16 shows the correlation between the nonadditive energies obtained from Eq. (4.21) and those resulting from the SCF calculation with the extended basis sets. Since the empirical values of the anisotropies of the bond polarizabilities are less reliable than the average polarizabilities, α, the nonadditive energies have also been calculated assuming δ_{OH} = 0 and keeping the same value of the average polarizability α_{OH} = 4.39 a.u. The results are seen to be only slightly worse than those obtained with $\delta_{OH} \neq 0$. Thus any uncertainty with regard to the anisotropy of the bond polarizability hardly affects our final results. The anistropy of the polarizability for the whole molecule is, of course, important but this is taken care of by splitting the total polarizability into the bond polarizabilities.

It is certainly gratifying to see that the nonadditive interaction energy for the water trimer calculated in the extremely simple way using the atomic point charges and bond polarizabilities agrees so remarkably well with that obtained in the extended SCF calculations. The agreement may even seem somewhat surprising since, obviously, the induction energy resulting from Eqs. (4.21) to (4.23) does not contain all the nonadditive components of the total interaction energy (137).

In conclusion for distances of chemical interest the nonadditive interaction in $(H_2O)_3$, that is, between polar molecules, has been found to be fairly small. The relative values of the nonadditive contribution to the interaction energy of $(H_2O)_3$ are smaller than in the case of ionic-polar systems, such as $Li^+...F^-...H_2O$ (132), or beryllium clusters (138). They are, however, considerably larger than for clusters of rare gas atoms (139).

For intermolecular distances close to the energy minimum, or larger, the long-range induction energy has been shown to give the dominant contribution to the nonadditive energy. For most configurations reliable values of the nonadditive energy can be obtained using minimal basis sets and the counterpoise method, or even a London-type formula which relates the induction energy to the bond polarizabilities and to point charges localized on atoms.

Let us now consider another system composed of two molecules and one ion. We have selected to analyze the lithium ion interacting with two water molecules (140). For simplicity in notation, in referring to the Li^+-$(H_2O)_2$ system, we shall designate the two water molecules as W_1 and W_2, the corresponding oxygen atoms as O_1 and O_2 and we shall refer to the plain defined by O_1, O_2 and Li^+ as "reference plane." Recently, Clementi, et al. (141) have considered over two hundred different configurations for the $Li^+W_1W_2$ system, previously reported (140). In all configurations a constant value of 3.55 a.u. for the intermolecular separation (O_1-Li^+) was used; this value corresponds to the minimum in the Li^+-H_2O system (68). Geometrical variables $R(O_2$-$Li^+)$ and the angle ϕ defined as O_1-Li^+-O_2 were selected (see Figure 17).

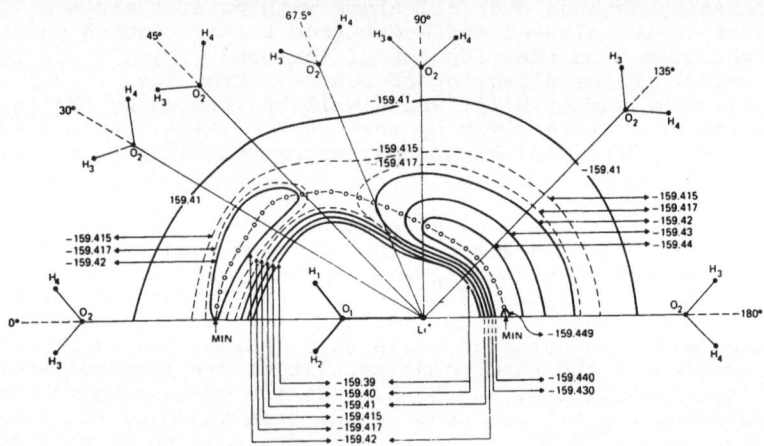

Figure 17. Hartree-Fock energy surface of the planner $Li^+(H_2O)_2$ in type 1 complexes. Total energies in a.u.

The configurations analyzed can be grouped into four types. In type 1 the molecular planes of W_1 and W_2 coincide with the reference plane; in type 2 only the plane of W_1 coincides with the reference plane, whereas for W_2 the molecular plane is perpendicular to the reference plane; in type 3 the role of W_1 and W_2 is inverted and in type 4 both molecular planes are perpendicular to the reference plane. The basis sets used in the computations are extended gaussian basis sets with polarization functions. The total interaction energy I for the Li^+-W_1-W_2 system is defined as

$$I = E(Li^+-W_1-W_2) - 2E(W) - E(Li^+) \qquad 4.24$$

with $E(W) = E(W_1) = E(W_2)$; this interaction energy has not been corrected for the basis set superposition error (95), since with the basis set adopted, the error is probably small relative to the quantities in discussion, that is, the three-body correction. The nonadditive correction to the two-body potentials is given by ΔE defined as

$$\Delta E = I - E(Li^+-W_1) - E(Li^+-W_2) - E(W_1-W_2) \qquad 4.25$$

From the comparison between the value of I and the corresponding values of ΔE, the nonadditivity correction was found definitely not negligible and as large as 10 to 15% of the total interaction.

Nonadditivity of intermolecular interactions can be conveniently discussed using the language of perturbation theory. Thus the exchange-type contributions to the interaction energy are nonadditive in all orders of the pertubation theory. The first order polarization energy, that is, the electrostatic energy, is additive. In the second order the polarization energy is composed of two parts: the induction energy which is nonadditive, and the dispersion energy which is additive.

As for third-order contributions, we shall mention only the Axilrod-Teller (142) term, related to the dispersion interaction and not included in the SCF interaction energies. However, in the case of polar systems the Axilrod-Teller term is small and can be neglected. Thus, except for the second order exchange dispersion energy which is believed to be small, the nonadditive energy, as calculated in the SCF approach contains all nonadditive contributions to the interaction energy through the second order.

In the previous study of water trimers (137) it has been shown that for this system the nonadditive interaction energy can be reasonably well approximated by the nonadditivity of the induction energy due to the interaction of atomic point charges and induced bond dipoles. If for this system the nonadditive exchange effects are small, they are likely to be relatively still smaller for $Li^+(H_2O)_2$ where the induction energy is larger than in $(H_2O)_3$. Hence one may expect that the nonadditive contributions to the SCF interaction energy of $Li^+(H_2O)_2$ can be approximated by the nonadditivity of the induction energy which can be easily evaluated.

Let us treat each water molecule in $Li^+(H_2O)_2$ as a system of two bonds polarized by the atomic charges of another H_2O molecule and of the Li^+ ion. The Li^+ ion will be assumed to be unpolarizable. Thus the total induction energy can be expressed as

$$E^{(2)}_{ind} = E^{(2)}_{ind,A} + E^{(2)}_{ind,B} \qquad 4.26$$

where A and B denote the two H_2O molecules, respectively, and

$$E^{(2)}_{ind,K} = -\frac{1}{2} \sum_{\lambda \epsilon k} [\alpha^T_\lambda (\xi_\lambda \cdot \xi_\lambda) + \delta_\lambda (\xi_\lambda \cdot e_\lambda)^2] \qquad 4.27$$

where α^T_λ and δ_λ denote the transverse polarizability and its anistropy, of bond λ in the molecule K, ξ_λ is the electric field vector at the midpoint of bond λ

$$\xi_\lambda = \sum_\mu \frac{q_\mu}{R^3_{\lambda\mu}} R_{\lambda\mu} \qquad 4.28$$

where the summation extends over the Li^+ ion and over atoms of the water molecule different from K, q_μ is the point charge of atom μ; $R_{\lambda\mu}$ denotes the radius vector from atom μ to the midpoint of the bond λ and e_λ is the unit vector in the direction of bond λ. From Eqs. (4.26) to (4.28) the nonadditive component of $E^{(2)}_{ind}$ can easily be obtained. It is given by the part of $E^{(2)}_{ind}$ which depends on the product $q_\mu \cdot q_\nu$ with μ and ν denoting atoms one of which is Li^+ and another belongs to the water molecule different from K. Hence, the nonadditive interaction energy in the system X.A.B, with an unpolarizable ion X, is proportional to the charge q_X of the ion X.

In numerical calculations standard values of the bond polarizabilities have been used: $\alpha^T_{OH} = 3.91$ and $\delta_{OH} = 1.42$ a.u.(143). For the point charges in H_2O we have used the values $q_O = -0.682$ and $q_H = 0.341$ ob-

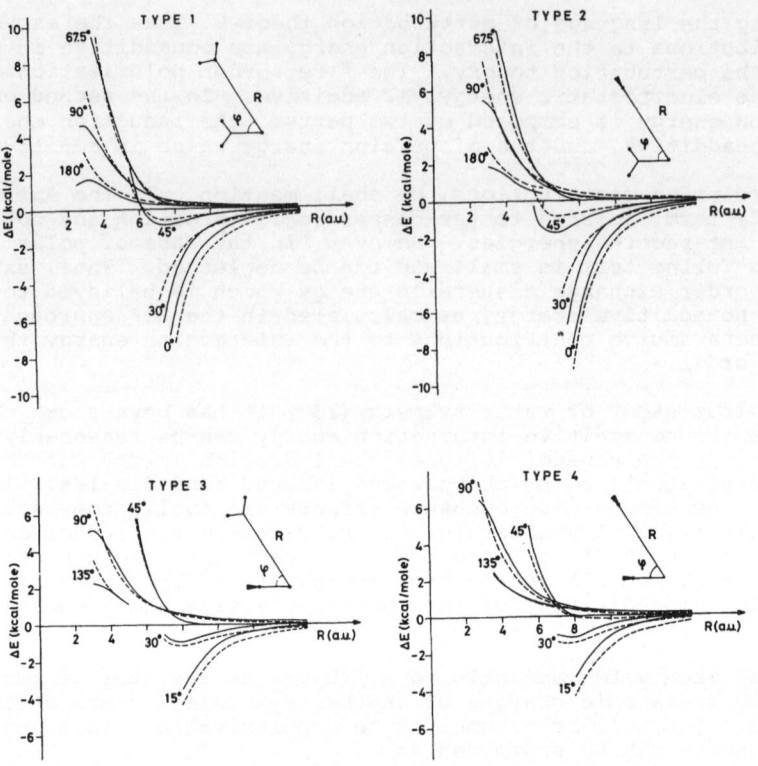

Figure 18. SCF and induction non-additivity for $Li^+(H_2O)_2$ in type 1, 2, 3 and 4 configurations.

tained using an extended basis set. The results for the four types of configurations of $Li^+-(H_2O)_2$ are shown in Figure 18 (inserts a, b, c and d) where the solid lines denote the results of SCF calculation and the broken ones have been obtained from Eqs. (4.26-4.28).

The agreement is seen to be very satisfactory. Appreciable differences between the two sets of results appear only for short distances where the nonadditivity of the first order exchange energy, related to the overlap effects, becomes important. Note, however, that the equilibrium Li...O distance in Li^+-H_2O is 3.55 a.u. Hence, the overlap effects not included in the second-order induction energy becomes important only for repulsive configurations of $Li^+-(H_2O)_2$.

The induction correction as defined by Eq. (4.26) describes the long and the intermediate-range behavior but breaks down at shorter distances. In this distance range the differences between the ΔE values from SCF and those obtained from Eq. (4.26) can be fitted by a correction term of the form

$$E_{sr} = A \exp\left\{-C_1 R(O_1-O_2) - C_2[R(Li^+-O_1)+R(Li^+-O_2)]\right\} \qquad 4.29$$

where the coefficient value of A, C_1 and C_2 are 1.52×10^5 a.u., 0.19 and 2.24 respectively and the distances R are expressed in a.u.

Let us rewrite equation (4.28) in the form

$$\xi_\Lambda \equiv \xi_{\lambda,K} = \sum_{J \neq K} \underset{\sim}{A}_{\Lambda J} \qquad 4.30$$

where

$$\underset{\sim}{A}_{\Lambda J} = \sum_{\sigma \in J} \frac{q_\sigma}{R_{\Lambda\sigma}^3} \underset{\sim}{R}_{\Lambda\sigma} \qquad 4.31$$

Here the subscript Λ refers to a bond in the K-th water molecule, the subscript J refers to another water molecule and/or to the Li^+ ion, and the subscript σ refers to the point-charge distribution therein. We shall write $\sum_\Lambda$ as a short-hand notation for $\sum_K \sum_{\lambda \in K}$.

For a general many-body situation $Li^+(H_2O)_n$ we have

$$\sum_\Lambda (\underset{\sim}{\xi}_\Lambda \cdot \underset{\sim}{\xi}_\Lambda) = \sum_\Lambda (\sum_{J \neq K} \underset{\sim}{A}_{\Lambda J})^2 = \sum_\Lambda \sum_{J \neq K} (\underset{\sim}{A}_{\Lambda J} \cdot \underset{\sim}{A}_{\Lambda J}) + \sum_{\Lambda J} \sum_L{'} (\underset{\sim}{A}_{\Lambda L} \cdot \underset{\sim}{A}_{\Lambda L}) \qquad 4.32$$

and

$$\sum_\Lambda (\underset{\sim}{\xi}_\Lambda \cdot \underset{\sim}{e}_\Lambda)^2 = \sum_\Lambda \sum_{J \neq K} (\underset{\sim}{A}_{\Lambda J} \cdot \underset{\sim}{e}_\Lambda)^2 = \sum_{\Lambda J} \sum_L{'} (\underset{\sim}{e}_\Lambda \cdot \underset{\sim}{A}_{\Lambda J})(\underset{\sim}{e}_\Lambda \cdot \underset{\sim}{A}_{\Lambda L}) \qquad 4.33$$

where the primes in the summation signs mean

$$I \neq K,\ L \neq J,\ J \neq K;\ \underset{\sim}{e}_\Lambda = \underset{\sim}{e}_{K\lambda}$$

refers to λ-th bond in the K-th water molecule.

We define the nonadditive induction energy for our many-body system by

$$E_{ind} = -\frac{1}{2} \left\{ \alpha^T [\sum_\Lambda (\underset{\sim}{\xi}_\Lambda \cdot \underset{\sim}{\xi}_\Lambda) - \sum_\Lambda \sum_{J \neq K} (\underset{\sim}{A}_{\Lambda J} \cdot \underset{\sim}{A}_{\Lambda J})] + \right.$$

$$\left. + \delta [\sum_\lambda \underset{\sim}{\xi}_\Lambda \cdot \underset{\sim}{e}_\Lambda)^2 - \sum_\Lambda \sum_{J \neq K} (\underset{\sim}{A}_{\Lambda J} \cdot \underset{\sim}{e}_\Lambda)^2] \right\} \qquad 4.34$$

As for the short-range correction term, we sum Eq. (4.29) over all pairs of oxygen atoms.

We conclude by stating that the nonadditivity correction, can be handled rather simply and reliably. The use of the generalized many-body correction given in Eq. (4.34) has been tested in a Monte-Carlo simulation of a lithium ion enclosed in a small cluster of molecules (141), yielding remarkably close agreement with accurate experimental data.

5.0 COMPLEXITY BECAUSE OF THE NUMBER OF COMPONENTS IN THE CHEMICAL SYSTEM

5.1 LIQUID WATER

In the past rare gases provided the testing ground for theories and simulations on the liquid state; today liquid water has assumed that role. The added complexity in the intermolecular interactions and the importanc of water as a solvent are two of the reasons for this shift. Pioneering studies in this direction have been presented by Baker, et al.(144) where Monte Carlo simulations based on empirical water-water potentials were considered. In the following we shall be mainly concerned with quantum-mechanically derived potentials.

After preliminary studies on a two-body Hartree-Fock potential(89a) and attempts to introduce dispersion corrections(89b), a more accurate two-body potential was derived(145) based on a configuration interaction stud on the dimers of water.

Were we to start such effort today, we would probably use an appropriate minimal basis set, the counterpoise method, and dispersion corrections as analyzed in the fourth chapter. At any rate an accurate potential is now available, and three-body corrections are also available. We refer elsewhere for a review on the water potential, small water clusters and liquid water(70). Here we recall that MC simulations on the liquid water(90) yield both X-rays and neutron-beams scattering intensities that are in notable agreement with experimental data as shown in Figure 19. The existing discrepancy, concerning the intensity of the first pick is due to the neglect of many-body corrections.

Even if fully expected, it is gratifying to note the steady and gradual improvements when we pass from the two-body Hartree-Fock potential to the correlated two-body potential; inclusion of three-body correction is under study. These results can lead to two different points of view, each one reasonable. On one hand, one might conclude that empirically derived potentials are rapidly becoming obsolete and one should rely more and more on quantum chemically derived potentials. The alternative view is that we should retain empirical potentials at least for the many-body correction.

The accurate two-body potential for the water-water interaction(145) has been used by Beveridge and his group of researchers(146,147). Of particular interest is the simulation of entropy(147). The computed entropy is -14.33 ± 0.03 to be compared with an observed value of -13.96 cal/deg mole. An alternative avenue for entropy simulations is the one considered by Romano and Singer(148). In view of the importance of the entropy parameter in solution studies, we shall briefly comment on entropy simulations.

In a canonical ensemble, the average value of a quantity F is defined as

where
$$<F> = \int F(Q) \exp[-\beta W(Q)] dQ / Z(Q) \qquad 5.1$$
$$Z(Q) = \int \exp[-\beta W(Q)] dQ.$$

For $\qquad F(Q) = \exp[\beta W(Q)] \qquad$ one obtains

$$<F> = \int dQ / Z(Q) \qquad 5.2$$

Unfortunately the above quantity cannot be obtained from MC simulations, since in the Markoff chain those configurations which heavily contribute

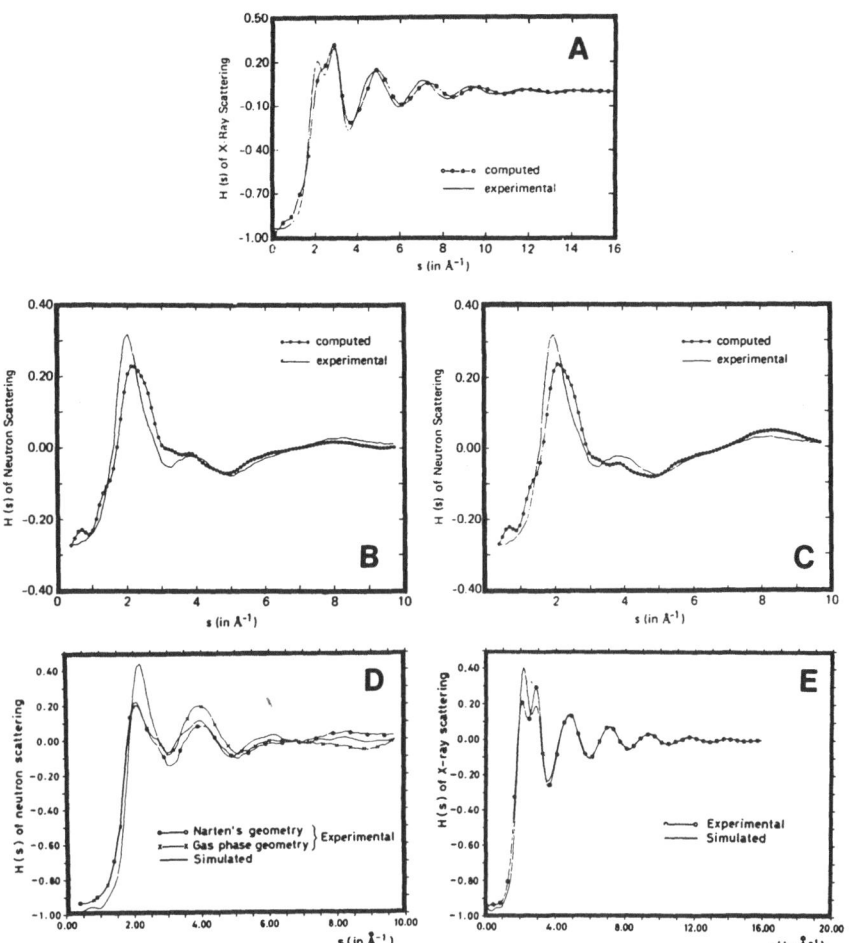

Figure 19. Scattering intensities (computed and experimentals) from X-rays (A and E) and neutron beams (B, C and D); Inserts A, B and C report on SCF+ perturbation potential; Inserts D and E refer to a C.I. potential.

to $\exp[-\beta W(Q)]$ have the smallest Boltzmann factor. As a consequence the entropy and thus the free energy of a system is obtained by alternative algorithms that can be classified as direct or indirect methods.

Among the indirect methods we recall, for example, those by Lebwohl and Lashar(149) and those by Torrie and Valleau(150,151,152). In the first one the Helmoltz function A for the N particle system is obtained either as

$$A(T_2,N,V)/T_2 - A(T_1,N,V)/T_1 = -\int_{T_1}^{T_2} (E/T^2)\, dT \qquad 5.3$$

or, alternatively, as

$$A(T,N,V_2) - A(T,N,V_1) = -\int_{V_1}^{V_2} P\, dV = -\int_{V_1}^{V_2} -\left(\frac{\partial A}{\partial V}\right)_{N,T} dV \qquad 5.4$$

In the second proposal, A is obtained by comparing the free energy of two systems (N,V,T,U) and (N,V,T,U_o), U being the two-body potential. Beveridge's work(147) makes use of these methods(5.3), that seems to present no convergence problems.

Among the direct methods, we recall the one proposed by Widom(153), that has been used, for example, in the work by Romano and Singer(148). As known, the chemical potential can be written as

$$\mu = (\partial/\partial N)(-KT \ln Z_N) \qquad 5.5$$

or, if we consider a monoatomic gas

$$\mu = (-3/2)KT \ln(2\pi mKT/h^2) + (\partial/\partial N) \ln\Theta_N \qquad 5.6$$

where

$$\Theta_N = (1/N!) \int \exp[-\beta W_N(Q)] dQ \qquad 5.7$$

For N sufficiently large one obtains:

$$(\partial/\partial N) \ln\Theta_N = \ln(\Theta_{N+1}/\Theta_N) \qquad 5.8$$

Let us increase the system from N to N+1 particles and let us impose the condition that the added particle experiences the field of the N particles but has not effect upon them. The added probe-particle is often referred to as "ghost particle". The potential on the probe-particle is

$$\Phi = \sum_{K=1}^{N} U(q_{N+1}, q_K) \qquad 5.9$$

If we designate as B the integrand

$$B = \exp(-\beta W_N) dx_1, \ldots, dx_N \qquad 5.10$$

we can write

$$\Theta_{N+1}/\Theta_N = (V/N)[\int \exp(-\beta\Phi)B / \int B] =$$
$$= (V/N) <\exp(-\beta\Phi)> \qquad 5.11$$

and we obtain

$$(\mu/KT) = (\mu_o/KT) - \ln<\exp(-\beta\Phi)> \qquad 5.12$$

where μ_o is the chemical potential for the system considered as an ideal gas.

The simulation for Br_2 and I_2 gases carried out by Romano and Singer(148) are very accurate, relative to experimental data. For condensed systems a possible drawback of the method is that the probe-particle will experience a very high replusion from the remaining particles in the system, and this fact brings about low numerical accuracy. A tentative solution is to consider the quantity $<\exp(-\beta\Phi)>$ as the numberical interpolation of a set values for $<\exp(-\beta\Phi)>$, obtained by an ideal compression and decompression of the simulated sample.

To conclude this section, by considering the above quoted papers and others, like those by Slanina(154) it seems clear that real chemical liquids rather than rare-gas liquids are now amenable to increasingly accurate simulations. There is, however, a tendency to attempt to define the structure of a given liquid, by proposing the model for the liquid in consideration. In our opinion, one should use much care in such definitions, since one of the notable aspects of a liquid is often the lack of one structure and the presence of many structures which can dynamically coexist in different subvolumes of the liquid, or at different time intervals.

5.2 ION WATER CLUSTERS: TWO BODY POTENTIALS

To study ionic solutions at the molecular level, ion-water and water-water potentials are basic prerequisite. In ion-water complexes, one of the main interaction is the coulombic interaction as pointed out by Scrocco and Salvetti in early 1950, the first quantum chemical computation reported on ion-water interactions. Accurate quantum chemical computations were performed starting in early 1970, at the Hartree-Fock level or at the C.I. level and with extended basis set (68,85,86,87,155). In the late 1970's subminimal and minimal basis sets have been published yielding, as expected, either very poor or poor binding energies (see Table 24).

As rather obvious, on the base of intuitive classical electrostatic reasoning a water molecule binds to a monoatomic anion via hydrogen bond formation, and to a monoatomic cation via utilization of the oxygen lone pairs. These different ways of binding for the water-ion complexes imply a different correlation energy correction in cation complexes relative to anion complexes, namely the correlation energy correction is expected to be more important for the anion complexes than for the cation complexes. As pointed out in the first chapter, in any discussion on the correlation energy correction it is basic to define a correct reference state. For singly-charged cations, I^+, the dissociation products of the complex

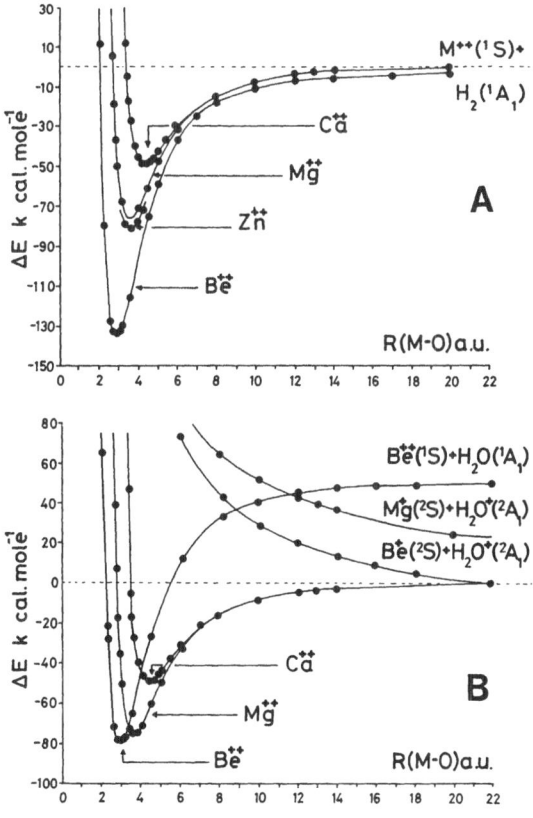

Figure 20. Double charged ions interacting with water: Above (A) at the Hartree-Fock level, below (B) at the C.I. level.

[I-H$_2$O]$^+$ are I$^+$ and H$_2$O. For doubly-charged cations, I^{2+}, the dissociations products are either I$^+$ and H$_2$O$^+$ or I^{2+} and H$_2$O. We recall that the process H$_2$O-->H$_2$O$^+$ costs between 12.6 e.v. and 14.5 e.v. (depending on the electronic state symmetry considered). We recall, in addition, that this energy can often be supplied by the process I^{2+}-->I$^+$ if the difference between second and first ionization potentials is sufficiently large. In Figure 20 we present SCF computed Binding energies for Be^{++}, Mg^{++}, Ca^{++} and Zn^{++} with a water molecule using a very extended basis set and the corresponding computations with C.I.(45). These computations have been performed as reported in reference(155) for the case of Be^{++}-H$_2$O. For a detailed discussion we refer elsewhere(155). Here, however, we add that for poliatomic singly-charged cations the possibility of the dissociation into I +H$_2$O$^+$ must be considered. For a review on Li$^+$, Na$^+$, K$^+$, Cl$^-$, F$^-$ complexing with water and on Monte Carlo simulations on water clusters containing either single ions or ions pairs, we refer elsewhere(70). One aspect, however, is recalled. In the cluster study, with either a single ion or an ion pair, the entire water cluster, composed of 200 water molecules, was found to be an hydrogen bonded network having a pole at the ion (or ions). Three-dimensional pictures of specific conformations of the ion-water cluster constantly revealed the existence of water-water filaments starting at the ion and radiating towards the surface, in single-water cluster, or bending towards the second ion, in ion-pair-clusters. These pictures did closely resemble the iron particle patterns obtained by a magnetic field, and suggested the possibility of describing aspects of ionic solutions with an effective two-body potential associated with the solvated ions and only implicitly including the water-water interaction.

5.3 IONIC SOLUTIONS: EFFECTIVE TWO BODY POTENTIALS

We can view an ionic solution either as composed of ions and solvent molecules, or as composed of solvated ions. In the latter case, the solvent does not appear explicitly. These two views, being models, can be used to describe different aspects of an ionic solution. In the first point of view, we follow the interaction of all the species composing the ionic solution, in the second model we think of the ionic solution as of macromolecules (the solvated ions) interacting among themselves. Depending on the concentration and temperature, the size of a given macromolecule fluctuates, with the condition, however, that the total number of solvent and solute molecules must remain constant. The interaction between the "macromolecules" must account for these fluctuations.

The fundamental characterization of the effective potentials among these macromolecules is obtained by considering the standard two-body potentials, previously discussed, and the pair correlation function obtained, for example, from Monte Carlo simulations. Let us consider an anion(A) and a cation(C); and the corresponding pair correlation functions (see top of Figure 21). The pair correlation functions inform us of the number of hydration shells (see middle region of Figure 21). Arbitrarily, we have selected the oxygen-cation and the oxygen-anion pair correlation functions to define the shell's radius. For a more detailed discussion we refer elsewhere(156). Our aim is to construct a two-body potential for the interaction between two solvated ions. An idealized form of such effective potentials is given at the bottom of Figure 21, designated as $E_1(r_1)$ and $E_2(r_2)$ for the cation and the anion, respectively.

With the definitions given in Table 33, it is easy to obtain $E_1(r_1)$ and $E_2(r_2)$, (see bottom of Figure 21). These potentials are assumed to be equal to the water-ion potentials for $0 \leq r_1 \leq r(c)$ and for $0 \leq r_2 \leq r(a)$ and to be analytically related to the water-ion potential in the regions

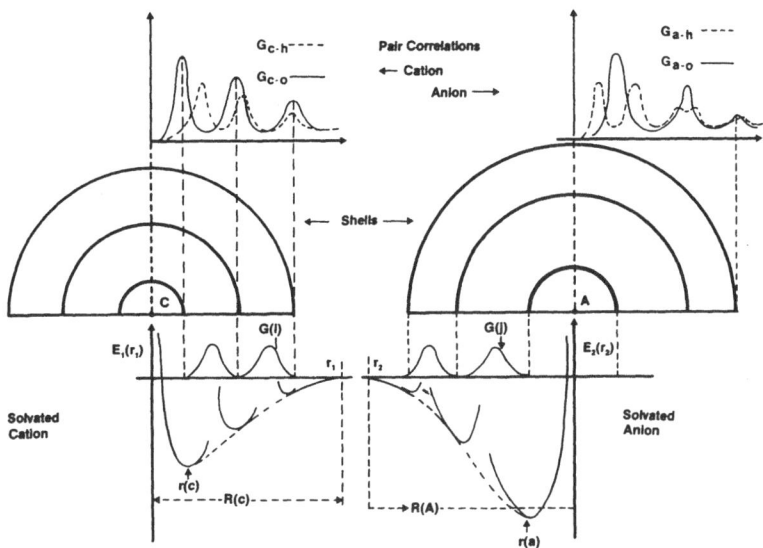

Figure 21. Solvated ion potentials: pair correlation functions (top) hydration shells (middle) and cation-anion potentials (bottom).

Table 33. Parameters for $E_1(r_1)$ and $E_2(r_2)$

```
r(c)  oxygen-cation, equilibrium distance in the complex H₂O-C.
r(a)  oxygen-anion, equilibrium distance in the complex H₂O-A.
n'(c) number of shells surrounding C at temperature T
n'(a) number of shells surrounding A at temperature T
n(c) = n'(c) + 1
n(a) = n'(a) + 1
r(w)  oxygen-oxygen distance in bulk water at temperature T
V(c)  potential energy for the complex H₂O-C
V(a)  potential energy for the complex H₂O-A
x(c)  distance oxygen-C (for V(c))
x(a)  distance oxygen-A (for V(a))
ρ(c)  value of x(c), where V(c) = KT
ρ(a)  value of x(a), where V(a) = KT
R₁ = r(c) + n(c) r(w)
R₂ = r(a) + n(a) r(w)
K(c) = (B(c)-r(c))/( (c)-r(c))
K(a) = (B(a)-r(a))/( (a)-r(c))
I_A(r₁) = value equal to V(c) at x(c) = r₁/K(c)
I_B(r₂) = value equal to V(a) at x(a) = r₂/K(a)
i = running index with values 0, ...n'(c)
j = index between 0 and n'(a)
G(i)  = gaussian function centered on r₁ = r(c) + (i+ 1/2)r(w)
G'(j) = gaussian function centered on r₂ = r(a) + (j+ 1/2)r(w)
```

$r(c) < r_1 < R_1$ and $r(a) < r_2 < R_2$. In the latter region the relations

$$\begin{cases} E_1(r_1) = I_A(r_1) + \exp(-\alpha r_1)\ G(i) \\ E_2(r_2) = I_B(r_2) + \exp(-\beta r_2)\ G'(i) \end{cases} \quad 5.13$$

are defined with the help of Table 33. The constants are specific for a given ion. The energy of a system composed of cations and anions and water molecules, can now be approximated by considering the interaction of the solvated ions, the repulsions due to the cation-cation and anion-anion interactions and for the solvent outside the solvation sphere of radius R_1 and R_2 (see Table 33) the average energy of bulk water. Let V be the volume of the ionic solution, $m(a)$ and $m(c)$ the number of anions and cations; the bulk water energy is proportional to

$$V = \sum_{i=1}^{m(a)} v_1(i) - \sum_{j=1}^{m(c)} v_2(j) \qquad 5.14$$

where

$$4/3\pi r(c)^3 \le v_1(i) \le 4/3\pi R_1^3 \qquad 5.15$$
$$4/3\pi r(a)^3 \le v_2(j) \le 4/3\pi R_1^3 \qquad 5.16$$

In the numerical simulation, the solvated ions can assume random positions within V; by analyses of the interionic distances, the volumes $v_1(i)$ and $v_2(j)$ are determined, with the help of the effective potentials $\bar{E}_1(r_1)$ and $E_2(r_2)$.

Before concluding, it is noted that the idea to use effective potentials to represent solvated ions is rather old; we refer to the papers by Bernal and Fowler(158), by Verwey(159) and by Morf and Simon(160), as examples of previous attempts.

5.4 IONIC SOLUTIONS: N-BODY CORRECTION

We have previously discussed techniques to obtain three and n-body interaction energies. Recently a Monte Carlo simulation on the systems $Li^+-(H_2O)_n$ has been performed by inclusion of many-body induction correction and short-range correction (see eq. 4.29 and 4.34). The computational results are analyzed in Table 34 and in Figure 22. The data reported indicates clearly the importance of the inclusion of many-body corrections. The number of solvent molecules is relatively small and therefore, it is somewhat premature to determine the increase in computer time caused by the inclusion of many-body effects; from our preliminary data it seems, however, that the time increase is within a factor of two.

5.5 ENERGY MAPS AND WATER STRUCTURE IN SOLUTIONS

We shall briefly comment on the relationships between the <u>isoenergy contour maps</u> obtained for the interaction of a single molecule of water with a biomolecule and the structural organization of many molecules of water for the biomolecule in solution, obtained by the Monte Carlo technique. The reader should keep in mind that the maps can be obtained for any two molecules, not only for a biomolecule and a water molecule. This comment has been added because we feel that the following is relevant to drug design, to intercalation problems in DNA and to a variety of chemical problems.

The isoenergy contour maps quantitatively represent the interaction energy of a single molecule of water at any position on a predetermined surface (generally a planar surface). More specifically, the maps are obtained by fixing the oxygen atom of a water molecule at the intersection point of a fine grid (covering the surface) and by optimizing the hydrogen-atom orientation in such a way as to obtain the minimum interaction energy between the water molecule and the biomolecule. The optimization is performed by using the previously discussed analytical potentials; since the latter yield interaction energies very near to ab initio computations, it follows that isoenergy maps are as nearly accurate as ab initio computations performed in the so called "supermolecule approach", and at a computational speed faster several hundreds to several thousands times. This gain in speed is the necessary requirement to perform statistical simulation, that obviously can not even be considered at the "supermolecule" level. Incidentally, we note that the term "supermolecule" is rather unfortunate; the term "whole complex" originally used for such computations, seems to us preferable(53). In general the

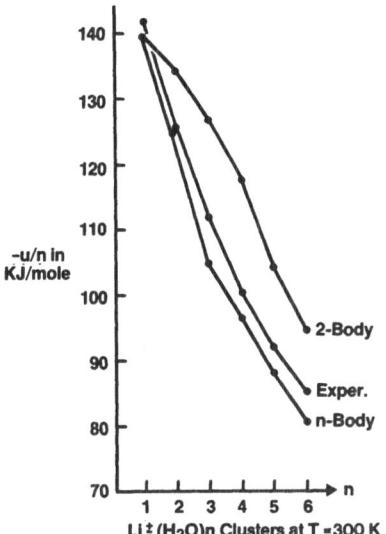

Figure 22. Monte Carlo entalpy experimental and simulated with 2-body and with n-body potentials.

Table 34. Li$^+$-(H$_2$O)$_n$ Clusters at 300K (all energies in K.J./mole)

n	a) $-U_2/n$	b) $-U_3/n$	c) $-\Delta H/n$	d) $-\Delta H\text{exp.}/n$
1	140.1	---	---	142.3
2	134.9	125.9	---	125.2
3	126.9	104.7	---	112.3
4	118.0	97.2	---	101.4
5	104.7	88.1	---	92.8
6	94.9	80.5	---	85.7
7	88.6	75.9	80.5	
8	81.5	70.8	74.7	
9	76.2	66.8	70.2	
10	71.1	64.7	67.0	
11	68.4	61.1	63.8	
12	65.2	58.5	60.9	
13	62.8	55.7	58.3	
14	61.5	53.6	56.5	
15	59.5	52.7	55.2	
16	59.1	51.9	54.5	
17	55.2	51.4	52.8	
18	53.3	50.4	51.4	
19	51.8	49.2	50.1	
20	50.5	47.2	48.4	
Deviation	±0.2	±0.4	±0.4	

a. Two-body potential
b. N-body potential (eqs. 4.29 and 4.34)
c. Recommended values
d. Experimental values from I.Dzidic and P. Kebarle, J.Phys.Chem. 74, 1466 (1970) and M. Arshadi, R. Yamdagni and P. Kebarle, J.Phys.Chem. 74, 1475 (1970)

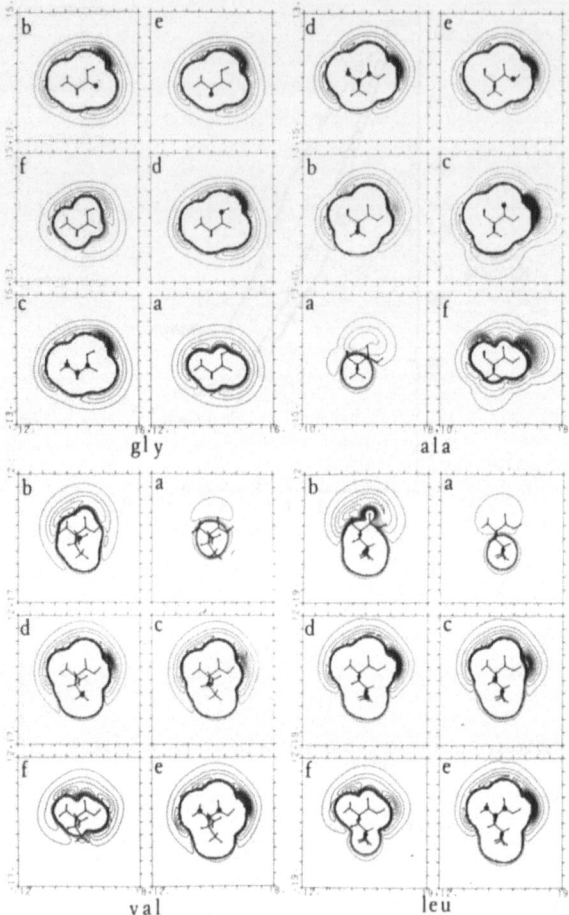

Figure 23. Iso-energy contour maps for Gly, Ala, Val and Leu.

interval between two successive grid points is 0.5-1.0 a.u. Graphical computer routines are used to connect with a contour line the position corresponding to the same value of the interaction energy water-biomolecule. For additional details, we refer to a previous work,(88,89 where isoenergy contour maps were used for solution studies. (The iso-energy maps should not be confused with the equipotential maps obtained by considering the electrostatic interaction between two molecules.)

In a contour map, we can distinguish four regions: (1) the molecular hard core, (2) the hydrophobic, (3) the hydrophilic and (4) the bulk water regions. The hard-core is characterized by positive (repulsive) contours exponentially decreasing from extremely high values to the zero value; in general one reports only the beginning of the repulsive region (the first few contours, including the zero-energy contour). These hard-core boundaries provide the shape of the molecule as seen by water; this shape may be different from our expectation, since we are often conditioned by the shape of molecules obtained by using X-rays rather than water as a probe. The hydrophobic region, if present, immediately

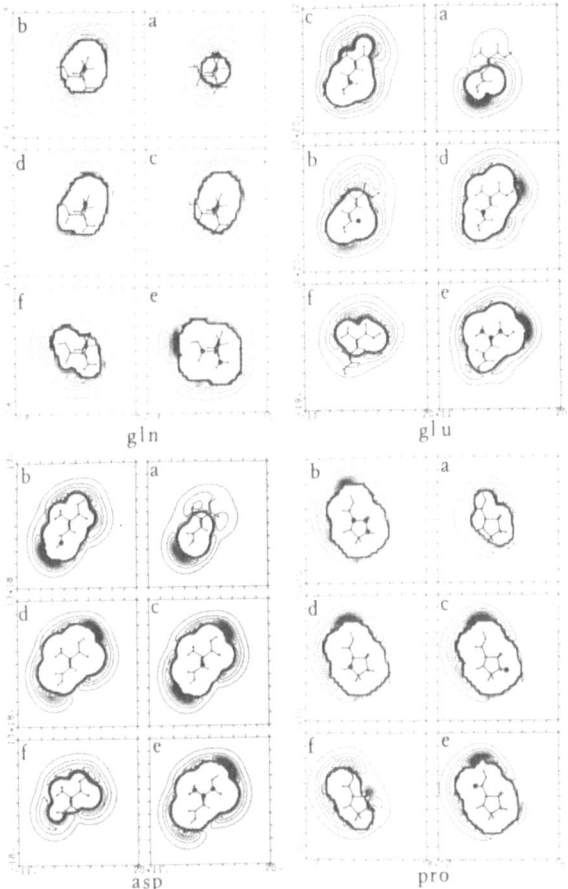

Figure 24. Iso-energy contour maps for Gln, Glu, Asp and Pro.

extends from the border of the hard-core region for about 4-5 A. In the absence of attractive contours, the bulk-water region starts beyond the hydrophobic region. Thus water penetrates from the bulk water region to the hydrophobic region that is constituted essentially by the first solvation shell bordering that part of the hard-core perimeter with hydrophobic character. The hydrophilic region is characterized by negative (attractive) contours and is delimited by the zero-energy contour of the hard-core and by a second zero-energy contour enveloping the set of negative contours (examples are provided in the next section). For the hydrophilic region, we distinguish several cases, based on two criteria: the depth of the minimum and width of the hydrophilic region near the minimum. The first factor mainly influences the orientation of a molecule of water; the second, mainly its position as described below. The hydrophilic region can be defined as "strong," "intermediate," and "weak," depending on the value of its attractive energy for a molecule of water. If the attraction is between 0 and 8 Kcal/mole, we call it weak; between 8 and 15 Kcal/mole, intermediate; more than 15 Kcal/mole, strong.

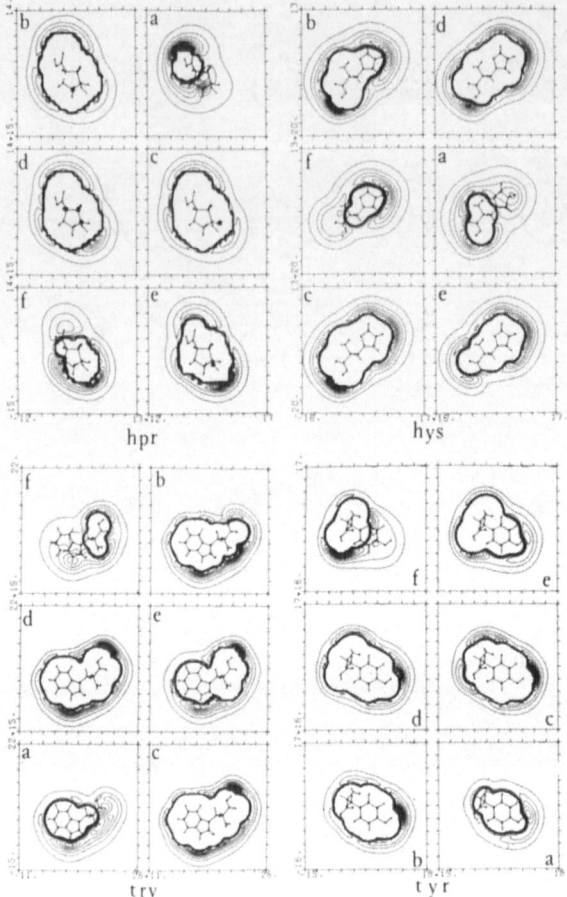

Figure 25. Iso-energy contour maps for Hpr, Hys, Try and Tyr.

In strong regions having a volume (width near the minimum) somewhat larger than the volume for one molecule of water (approximately a sphere of 3 A in diameter), the position and orientation in solution coincide with those obtained from the isoenergy contour map. In strong regions, but with a larger volume and well-defined minima (not shallow), the same conclusion holds. However, in strong regions again with a large volume but without well-defined minima, the orientation in solution is the one determined by the isoenergy maps, but the position depends on the steric condition determined by neighboring molecules of water.

In intermediate regions the above observations hold, although not fully, since in this case both positions and orientation are partly determined by neighboring water molecules.

In the weak regions, the position and the orientation always depend strongly on the neighboring water molecules.

A complementary way of analyzing the strong, intermediate, and weak hydrophilic regions is to consider the ratio between the value of the

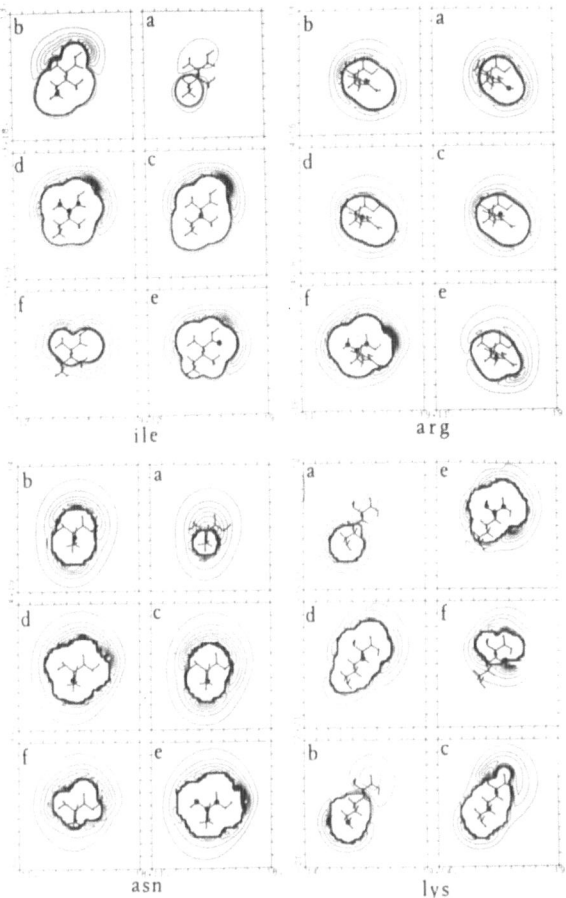

Figure 26. Iso-energy contour maps for Ile, Arg, Asn and Lys.

water-water interaction in bulk water and in the above three regions. In the weak case, the ratio is between a large positive value and 1; in the intermediate case, between 1 and 0.5; and in the strong case, between 0.5 and 0. The above criteria between the three situations for the hydrophilic regions are clearly somewhat rigid but are introduced to quantify the relationship existing between isoenergy contour maps for a single molecule and the interaction energy of water in solution. Therefore, the contour energy maps do not only provide the value of the interaction energy and the optimal position and orientation of a single molecule of water interacting with the biomolecule, but, in addition, allow one to infer with some reliability the position, orientation, and water-water interaction energy for the many water molecules present in solutions.

In Figures 23 to 28, we provide iso-energy contour maps for the naturally occurring amino acids in the neutral form and for a polyglycine. These maps have been obtained from the analytical atom-atom potentials, previously described(79,80,81,82). Some of the maps have been previously

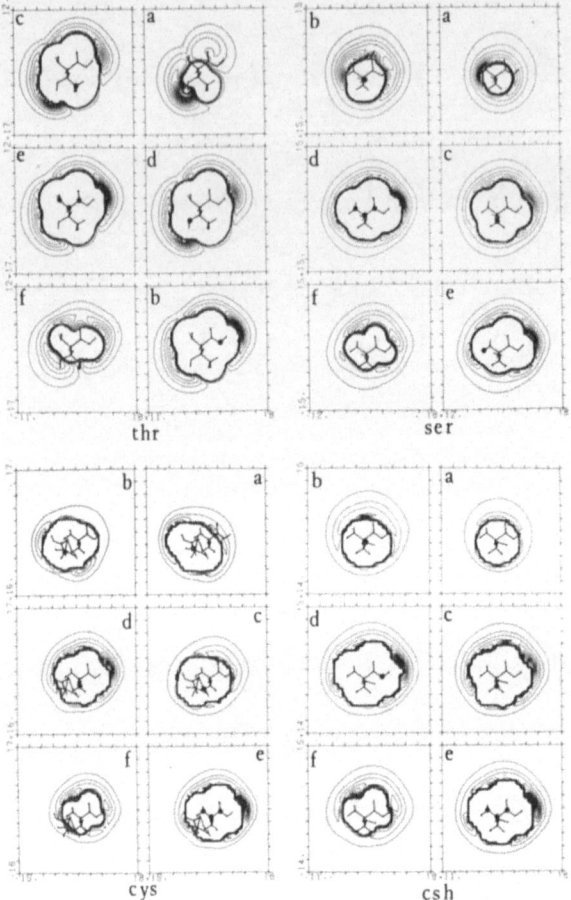

Figure 27. Iso-energy contour maps for Thr, Ser, Cys and Csh.

reported, but the full details have been available only in a series of a technical reports,Istituto G. Donegani, Montedison, Novara, Italy, 1975. For each amino acids <u>six parallel planes</u> have been considered (inserts a, b, c, d, e and f). For each amino acid <u>one plane</u> - called "reference plane" - contains three atoms of the $H_2N-CH-CO$-unit, namely the nitrogen, the $C(\alpha)$ and the C' atoms (see for example glycine in Figure 23 at insert "C"); such atoms are designated by a full dot. The remaining planes are either above or below the reference plane, and if an atom is contained in the plane a full dot is used to mark such atom. The contour to contour interval is 1.0 Kcal/mole. The first few contours (from the inside to the outside) are repulsive and very near one another and delimite the hard core region. The <u>outmost</u> contour is the zero energy contour; <u>when</u> such contour does not coincide with the hard core region (case of hydrophobicity), the contours between the outmost one and the hard core one, define the hydrophylic regions.

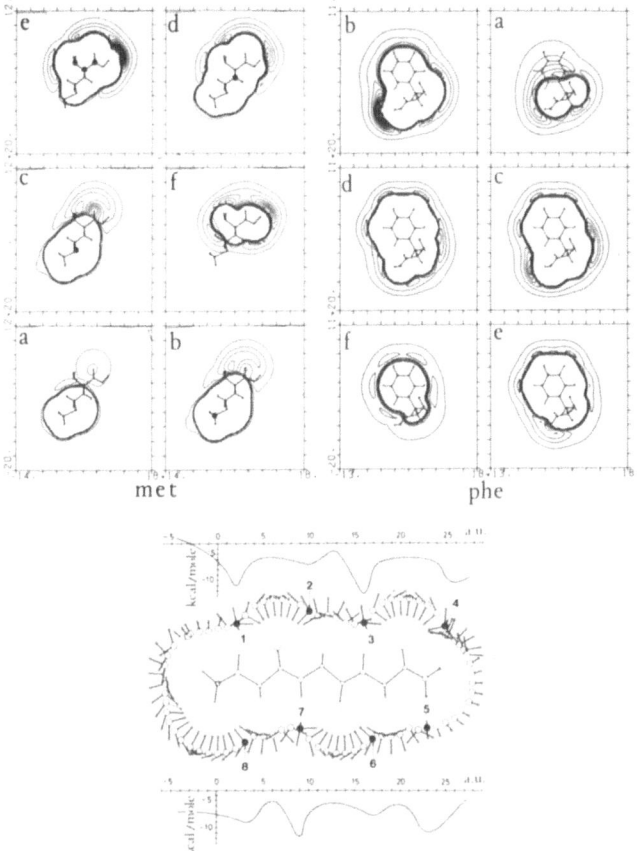

Figure 28. Iso-energy contour maps for Met and Phe. Bottom: water optimal orientation around polyglycine (Reference 32).

5.6 MONTE CARLO SIMULATION OF THE INTERACTION BETWEEN GLYCINE AND THE CORRESPONDING ZWITTERION

In recent years a number of calculations have appeared, dealing with solvation of molecules of biological interest; in general, the solvent is approximately represented by explicitly considering only one solvent (usually water) molecule. Thus, strictly speaking, these calculations correspond to the study of a two-body system at 0 K and in vacuo.

The implicit assumption is that solvent-solvent interactions can be ignored; this hypothesis can be reformulated as follows: it is known that in bulk water, the interaction energy amounts to about 36 kJ mol^{-1} and, on average, each molecule is tetrahedrally coordinated to four other water molecules, two of which form hydrogen bonds along the O-H bonds and the other two interacting with the oxygen lone pairs. If the solute-solvent interaction is greater than the solvent-solvent interaction by at least a factor of two, then the latter can be treated as a first-order correction. This is what could be expected for strongly polar groups, such as -NH$_3^+$ and -COO$^-$ in amino acid zwitterions: however, many biological molecules are composed not only by hydrophilic but also by hydrophobic groups, in the vicinity of which the water-water interaction is expected to be present even in the zero-order approximation. We are confronted with the tacit use of a hypothesis which might, possibly, be more acceptable in some regions of the biomolecule than in others.

The use of the Monte Carlo method is rather expensive in terms of computer time when realistic potentials are used. In order to perform a reasonably accurate test of our tacitly assumed working hypothesis, we have selected two biomolecules, suffiently small to allow a careful Monte Carlo analysis; they are neutral glycine and the glycine zwitterion(161). This choice was influenced by the presence of two hydrophilic (-NH$_2$ and -COOH or -NH$_3^+$ and -COO$^-$) and one hydrophobic (-CH$_2$-) groups. Moreover, the hydrophilic attraction in the zwitterion is about three times as large as the water-water interaction.

We shall attempt to obtain a detailed picture of the special arrangement of the water molecules surrounding the two glycines, and also of the special variation of their energies.

Monte Carlo calculations are carried out for a cluster consisting of the named amino acid and two hundred water molecules; all the relevant interaction potentials had been obtained by means of quantum mechanical calculations.

Table 35. Atomic Coordinates (Å), Classes And Net Charges For Glycine (a)

Atom	Class	x	y	z	Net Charge
O(2)	9	+2.131	-1.102	-0.377	-0.503
O(1)	10	2.096	1.058	0.348	-0.378
N	11	-0.549	-1.370	0.000	-0.548
C'	5	1.526	0.000	0.000	0.458
C(α)	8	0.000	0.000	0.000	-0.329
H(1)	1	-1.602	-1.374	0.044	0.241
H(2)	1	-0.249	-1.903	-0.836	0.254
H(1α)	2	-0.351	0.514	-0.895	0.181
H(2α)	2	-0.381	0.536	0.068	0.223
H—O(2)	4	3.068	-0.947	-0.337	0.398

Note: (a) Data taken from Reference 79.

Table 36. Atomic Coordinates (Å), Classes And Net Charges For Glycine Zwitterion (a)

Atom	Class	x	y	z	Net Charge
C(α)	8	0.000	0.000	0.000	-0.325
C'	32	1.530	0.000	0.000	0.433
O(1)	30	2.107	-1.109	0.000	-0.584
O(2)	30	2.107	1.109	0.000	-0.530
N	31	-0.493	-1.395	0.000	-0.662
H(1α)	2	-0.367	0.519	0.898	0.215
H(2α)	2	-0.367	0.519	0.898	0.215
H(1)	29	0.308	-2.043	0.000	0.438
H(2)	29	-1.066	-1.557	0.841	0.401
H(3)	29	-1.066	-1.557	0.841	0.401

Note: (a) Data taken from Reference 79.

Table 37. Energies Around The Hydrophilic Sites (a)

r,Å	Neutral Glycine (b)				Glycine Zwitterion (b)			
	N/W	N/A	C'/W	C'/A	N/W	N/A	C'/W	C'/A
2	—	—	—	—	—	—	—	—
3	-31	-11	—	—	-26	-31	-14	-50
4	-33	-3	-34	-9	-21	-60	-15	-54
5	-33	0	-36	-2	-35	-0	-32	-7
6	-36	0	-36	0	-33	0	-30	-1
7	-32	0	-32	0	-34	0	-38	0

Notes: (a) Estimated uncertainty ±1 kJ mol^{-1}.
(b) The interaction region is represented by the left hand symbol N or C'; the interacting species is indicated by the right hand symbol W (water) or A (amino acid). To allow a proper comparisoin the potential energy of a water molecule, due to the interaction with all the others is divided by 2.

The molecular geometries for the solute are given in Tables 35 and 36, together with additional data such as computed net charge and "class" for each atom. The water-water interaction potential was obtained from CI calculations and fitted to the following expression(145):

$$\Phi_{ww} = \frac{Q^2}{4\pi\varepsilon_0}\left[\left(\frac{1}{r_{13}}+\frac{1}{r_{23}}+\frac{1}{r_{14}}+\frac{1}{r_{24}}\right)+\frac{4}{r_{78}}+\right.$$

$$\left. -2\left(\frac{1}{r_{28}}+\frac{1}{r_{18}}+\frac{1}{r_{37}}+\frac{1}{r_{47}}\right)\right] + a_1\exp(-b_1 r_{56}) +$$

$$+ a_2[\exp(-b_2 r_{13})+\exp(-b_2 r_{14})+\exp(-b_2 r_{23})+\exp(-b_2 r_{24})] +$$

$$+ a_3[\exp(-b_3 r_{16})+\exp(-b_3 r_{26})+\exp(-b_3 r_{35})+\exp(-b_3 r_{45})] +$$

$$- a_4[\exp(-b_4 r_{16})+\exp(-b_4 r_{26})+\exp(-b_4 r_{35})+\exp(-b_4 r_{45})]$$

5.17

Here Q is a positive charge situated on each hydrogen atom. Hydrogen atoms in one molecule are labelled 1 and 2, and those in the other molecule are labelled 3 and 4; the two oxygen atoms are labelled 5 and 6, respectively, the label 7 refers to a point on the molecular C_{2v} symmetry axis and 0.2677 A away from the oxygen atom, where the charge $-2Q$ is situated; the subscript 8 is similarly defined in the second water molecule; ε_0 is the dielectric constant of vacuum, the molecular geometry was taken from reference 162, namely r(O-H) = 0.9572 A and H-O-H = 104.5°.

The amino acid-water interaction potential for neutral and zwitterion glycine was obtained in the SCF-LCAO-MO approximation (79-84).

Calculations were carried out at T = 300 K and all molecules were treated as rigid; results for the potential energies are shown below (in kJ mol^{-1}):

	Neutral Molecule	Zwitterion
water-water	-30.3 ± 0.1	-30.1 ± 0.1
Amino acid-water	-53.5 ± 0.8	-415.0 ± 5.0

They are averages over 1,000,000 configurations, after 500,000 configurations for equilibration; subaverages were taken over macrosteps of 50,000 configurations.

As far as the water-water interaction is concerned, a superficial interpretation might indicate that there is no significant difference between the two situations; the Monte Carlo result for bulk water (with the same interaction potential) is 35.6 ± 0.6 kJ mol^{-1}. On the other hand, the zwitterion appears, as expected, to be strongly favoured with respect to the neutral molecule.

A more refined interpretation of the energy results should allow for the fact that the water molecules sufficiently close to the solute are likely to interact with it to a significant extent, (depending on the solute and the atoms they are close to) and, on the other hand, water molecules which are sufficiently far from it will not be affected by the solute, but rather by surface effects.

For solvent molecules close to a hydrophilic group, one would expect that the stronger the solute-solvent attraction, the less important the solvent-solvent interaction, and this may well lead to non-optimum water-water separations and orientations, and, thus, to a water-water interaction energy smaller than in bulk water. On the other hand, water molecules should be pushed away from hydrophobic sites, but in our case a comparatively small hydrophobic group is sandwiched between two hydrophilic ones, and its effect is likely to be moderate or even quenched.

We considered two regions around the hydrophilic sites (see next section for their accurate definition) and, for each water molecule which happened to be in either region, we calculated the interaction energy with both all other solvent molecules and with the whole amino acid; for each of these water molecules we also calculated the distance of its oxygen atom from the appropriate "centre" (that is, N for the $-NH_2$ or $-NH_3^+$ group and C' for the -COOH or COO^- group) and, by averaging, we obtained the data shown in Table 37.

In Table 37 the water-water energy mostly increases with increasing distance. The perturbation induced by the hydrophilic groups (as measured by the amino acid-water interaction) is rather moderate for neutral glycine and rather marked with zwitterion. These results are expected to be of rather general validity for molecules in water solution and are of special interest for the study of solvent interactions in proteins and enzymes. Since the water-amino acid interaction drops sharply at 5 A, one can roughly estimate the thickness of the perturbed layer to be of one water molecule; these results are confirmed by structural analysis (see below).

The analysis in the vicinity of the hydrophobic group is remarkably more difficult, since the CH_2 region is sandwiched between two hydrophilic regions, and therefore the arbitrariness in the definition of a border rather severe. We plan to analyze this point with other amino acids, where the hydrophobic regions can be more easily identified.

Neither the amino acid nor the water molecules are point particles, and, thus one should calculate a generalized radial distribution function allowing for angular coordinates. However, this makes it difficult to visualize the physical meaning; moreover, different atoms in the amino acid "see" different water environments.

We decided to calculate radial distribution (RDF's) and orientational correlation functions (OCF's, to be defined below) around selected atoms and restricted to suitably chosen half-spaces.

We calculated RDF's and OCF's around the N and C' atoms. Whereas it is easy to define regions around the N and C' centres where water molecules are mostly acted upon by the corresponding groups, the solvation of the CH_2 group is more likely to be significantly affected by the two neighboring groups; we therefore decided to calculate two sets of RDF's and OCF's around the $C(\alpha)$ atom, over the two half spaces defined as follows: let one semiplane be defined by the z axis and the positive x axis, and another by the z axis and the $N-C(\alpha)$ bond. As for the OCF's of water dipoles, let us consider an atom s in the solute molecules, with coordinate vector $\underline{r}_s$, let $\underline{r}'$ be the coordinate vector of the oxygen atom in a solvent molecule, and u be the unit vector defining the orientation of its dipole (that is, the orientation of its C_{2v} symmetry axis). We define

$$F_s(r) = \langle \frac{\underline{r} \cdot \underline{u}}{r} \rangle ; \quad \underline{r} = \underline{r}' - \underline{r}_s \qquad 5.18$$

as a function of r.

RDF's and OCT's are given in Figures 29 and 30 and the analysis of the information provided by them can be supplemented by considerations obtained from the isoenergy contour maps both for neutral glycine and the zwitterion.

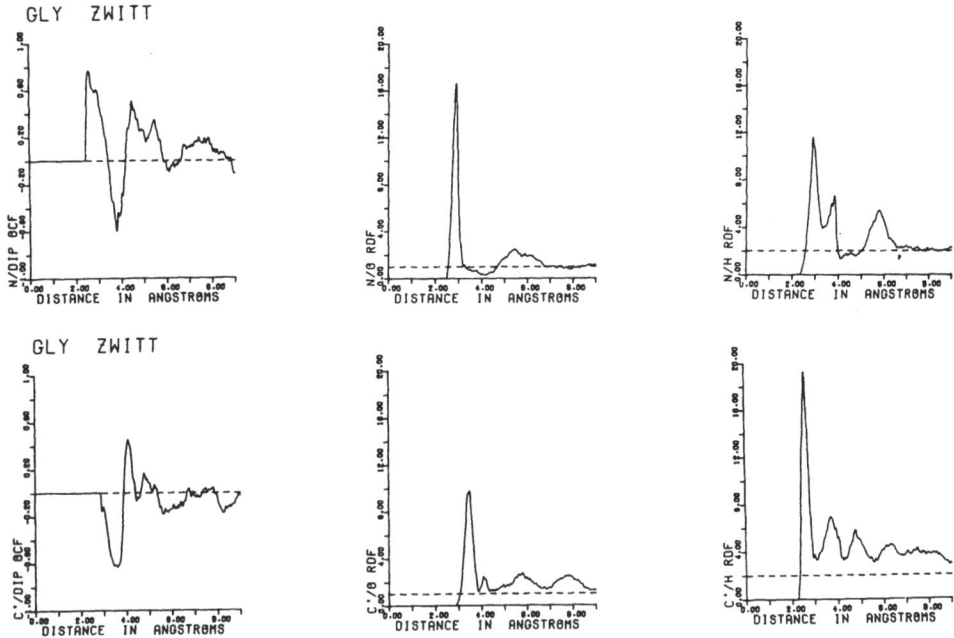

Figure 29. RFD's and OCF's for glycine zwitterion near N and C' atoms.

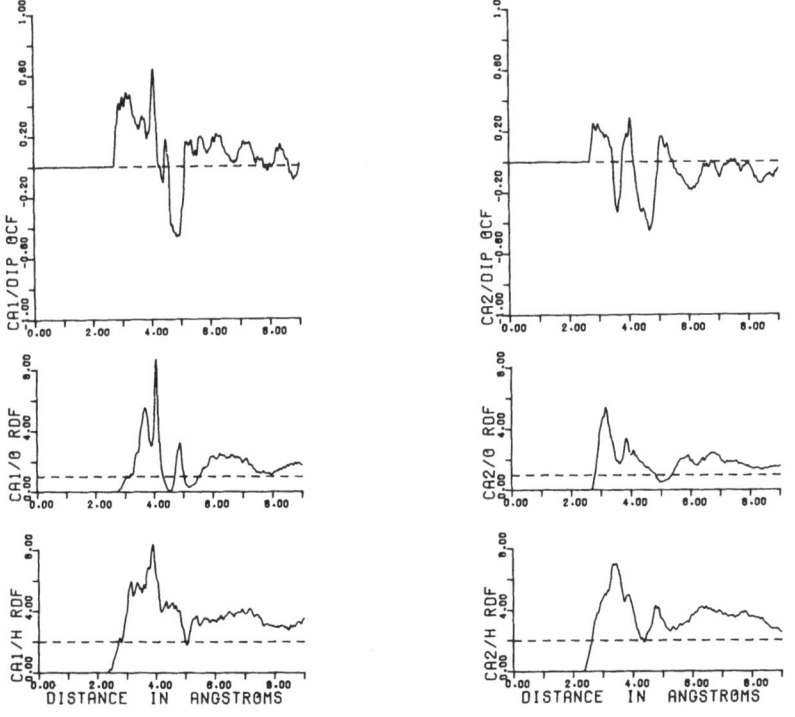

Figure 30. RDF's and OCF's for glycine and zwitterion (see text).

Let us compare the N sites; for the zwitterion (see Figure 29) the N/O
RDF shows a sharp peak at 2.9 A and a weaker, broader, one between 5 and
6 A; the N/H RDF has peaks at 2.9, 3.7 and 5.7 A: this site possesses
two solvation shells, the nearer one with a well defined orientation, and
the outer one with more rotational freedom and a less define orientation.
This picture is also supported by the dipole moment OCF (N/DIP OCF):
notice that the negative peak at 3.9 A has little significance, since the
N/O RDF shows a very small probability of finding oxygen atoms in this
region.

For neutral glycine the statistics is less clear, possibly owing to interference by the $-CH_2-$ group. However, there is a peak in the N/O RDF at
3 A, and a corresponding hydrogen peak at 2.1 A. Comparison between the
two N/DIP OCF's suggests that the orientational ordering of water molecules induced by the N site extends to greater distance for the zwitterion than for the neutral molecules.

We now consider the C' sites. For the zwitterion (see Figure 30) the
hydrogen peaks at 2.6 and 3.7 A can be associated with the oxygen peak
at 3.5A, indicating a hydrogen bond. There are other peaks, both for
hydrogen and oxygen atoms, but there seems to be no simple way of associating them, and also the C'/DIP OCF is difficult to interpret. This is
probably linked with the existance of a narrow minimum region surrounding
the COO^- site. In neutral glycine there is a main oxygen peak at 3.7 A
and two corresponding H peaks at 3.4 and 4.4 A; we have nearly the same
situation as for the zwitterion, but with less pronounced structure and
with water oxygens slightly farther from the C'.

About the number of solvation shells, we can conclude that around the N
and C' sites there are more than one shell in the zwitterion and certainly one for the neutral molecule; as for the hydrophilic group $-CH_2-$
we do not even assign one solvation shell, pending further analysis.

Thus, the hidden hypothesis at the basis of studies where a two-body system at 0 K is considered, receives some support; it is acceptable for
strongly hydrophilic regions, but does not appear reliable for weakly
hydrophilic or hydrophobic regions. An intermediate model(??), where
water molecules are added step-wise, starting from the most attracted one
and moving to the next, seems to be somewhat sounder, despite the drawback of neglecting temperature considerations and statistical reliability.

5.7 SERINE AND THE CORRESPONDING ZWITTERION

Previously we carried out Monte Carlo calculations on clusters consisting
of a glycine molecule, both in neutral and zwitterionic form, and 200
water molecules, using two-body interaction potentials obtained by quantum mechanical calculations; one of the aims of that calculation was to
assess the reliability of extrapolation to the many-body, finite-
temperature situation typical of a solution, on the basis of a two-body
zero-temperature potential, that is one solute and one water molecule.

We concluded that such an approximation might make limited sense in the
neighbourhood of strongly hydrophilic groups, but none for weakly hydrophilic or hydrophobic ones (see Section 5.5).

Here we extend our analysis to serine, and Monte Carlo simulate clusters
consisting of one serine molecule and 250 water molecules(163). Serine
was considered both as a neutral molecule (in the "crystallographic"(79a)
conformation) and as a zwitterion(79b). Moreover, in order to gain some
insight into conformational effects, we carried out calculations on two
conformers of the zwitterion, that is, the crystallographic one(79a) and

another conformation, among those examined in reference(79b) and found to be the stablest.

For the sake of simplicity the neutral molecule, the crystallographic and the non-crystallographic conformations of the zwitterion will be referred to as S1 (Table 18 in Reference(79a)), S2 (Reference(164)) and S3 (SERZI2 in Reference(79b)), respectively.

The three molecules are shown in Figure 31; there, and in all our calculations the reference frame was chosen so that the three atoms N, C_A and the carboxylic C' define the plane z=0, the C_A atom is the origin, C' belongs to the x-axis and the alcoholic C_B atom belongs to the half-space x>0. The plane x=0 will be called "molecular plane".

Calculations were carried out at T=300 K; results for the potential energies are shown below, in kJ.mol^{-1}, together with the corresponding results for glycine:

		S1	S2	S3
$U(w,w)$	serine	-30.6 ± 0.1	-30.4 ± 0.1	-30.8 ± 0.1
$U(w,s)$		-176.0 ± 3.0	-427.0 ± 5.0	-354.0 ± 4.0

		Neutral	Zwitterion
$U(w,w)$	glycine	-30.3 ± 0.1	-30.1 ± 0.1
$U(w,s)$		-53.5 ± 0.8	-415.0 ± 5.0

The figures quoted here are averages over 1 000 000 configurations after 500,000 configurations for equilibration; subaverages were taken over macrosteps consisting of 50,000 configurations. Two very simple conclusions can be drawn from these figures:

1. The zwitterion is favoured over the neutral molecule although less so than with glycine.

2. Recalling that the Monte Carlo result for the energy in bulk water (with the same interaction potential) is $35.6\pm.6$ kJ.mol^{-1}(90), the results on water-water interaction energy, both for glycines and serines, suggest that, whatever the solute and its immediate solvation, the energy contribution by more distant water molecules is predominant; to put it in another way, the numbers of water molecules we have been using are sufficient to saturate the solute-solvent interaction.

In order to obtain some information on the special variation of average interaction energies around the solute, we proceeded in the following way. The N,C_B,C' and C_A atoms were chosen as solvation centres; for N, C_B and C' the solvation region is the semispace containing the relevant atom and delimited by the orthogonal bisector of the segment joining C_A with the named atom; the solvation region for the C_A is defined as that part of space which does not belong to any other solvation region. These definitions are no doubt arbitrary, yet they are based on the reasonable assumption that water molecules in the neighbourhood of an N or C' centre will be mostly affected by the corresponding group; on the other hand, the C_A-H is a hydrophobic group sandwiched between two hydrophilic ones, and the arbitrariness in defining a border is much more severe. A somewhat different difficulty arises with the -CH_2OH group, which contains both hydrophilic and hydrophobic interaction centres and, moreover, is likely to interfere with the two hydrophilic ones. For each solvation region we worked out histograms of the average interaction energies (both with all other solvent molecules and with the solute) of water molecules belonging to it, as functions of the distances from the relevant centres. Some conclusions can be drawn (see Tables 1, 2 and 3 of Reference(163) for details):

1. In all solvation regions of S1 there is, at most, a moderate perturbation of the mutual organization of water molecules, as revealed by mean water-water energies smaller in magnitude than, say, 30 kj.mol^{-1}. Such a perturbation is more pronounced in the hydrophilic regions of the zwitterion, and is particularly profound in the C'region of S2, but not for S3.

2. For the N solvation region, the solvent-solute interaction is stronger for the zwitterion than for the neutral molecule; the results for the two zwitterion conformers are comparable, but the interaction appears to be longer ranged for S2 than for S3.

3. The differences between the two zwitterionic conformers can be rationalized as mainly due to differences in sterical interference between the hydrophobic CH_2 group and the two hydrophilic ones. Similarly, the fortunate situation in the C' region of the neutral molecule is probably connected with the fact that carboxylic proton sticks out farther from the CH_2 than the corresponding carboxylic oxygen atom in either conformation of the zwitterion.

Let us now analyze the water structure using the distribution functions. Since neither the solute nor the solvent molecules are point particles, one should, in principle, calculate a generalized distribution function allowing both for distances and angular coordinates, but then the visualization of the physical meaning for such a multidimensional hystogram is likely to be a hopeless task.

We could envisage two ways of circumventing this difficulty; one way consists of calculating radial distribution functions (RDFs) and dipole moment orientational correlation function around selected atoms in the solute molecule and restricted to suitably chosen half spaces, that is, for the N, C' and C_B regions we have already defined. We also defined two solvation regions for the C_A atoms, as for the glycine study. Unfortunately the corresponding plots did not show any significant feature, apart from the effect of neighbouring polar groups, and they are not reported here. Calculating RDFs actually implies smoothing the "granularity" of the solvent to an isotropic(though not homogeneous) medium. The other way of getting around visualization difficulties consists in calculating atomic probability density (PD) maps for suitably chosen "slices" of the volume containing our cluster, and is described in the following section. We examine below the structural details for each solvation region.

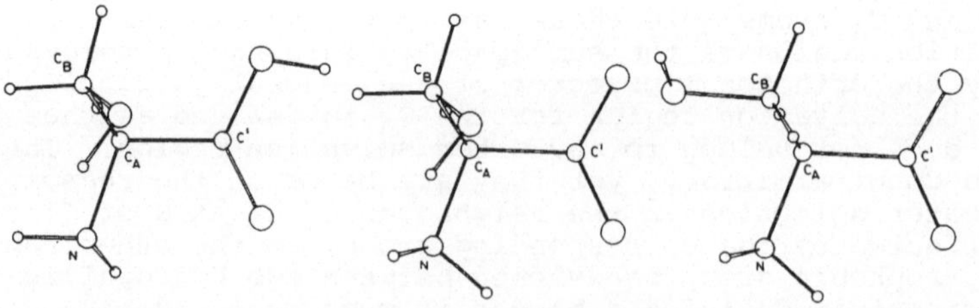

Figure 31. S1, S2, and S3 forms for serine as seen along the z - axis (ORTEP projection).

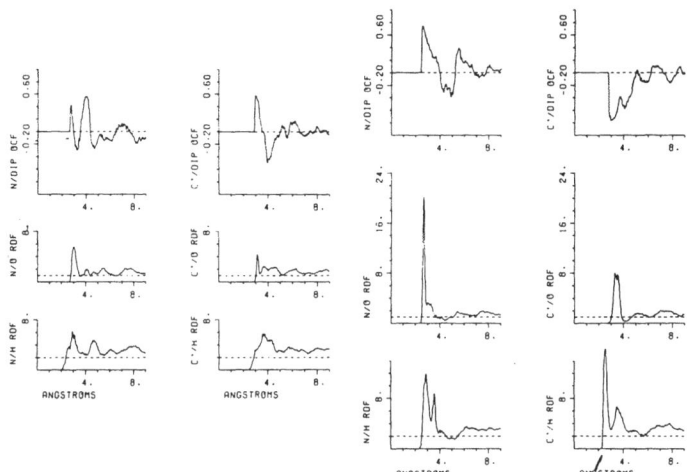

Figure 32. RDFs and OCFs for the N and C' regions of S1 (six inserts) to the left) and RDFs and OCFs for the N and C' regions of S2 (six inserts to the right).

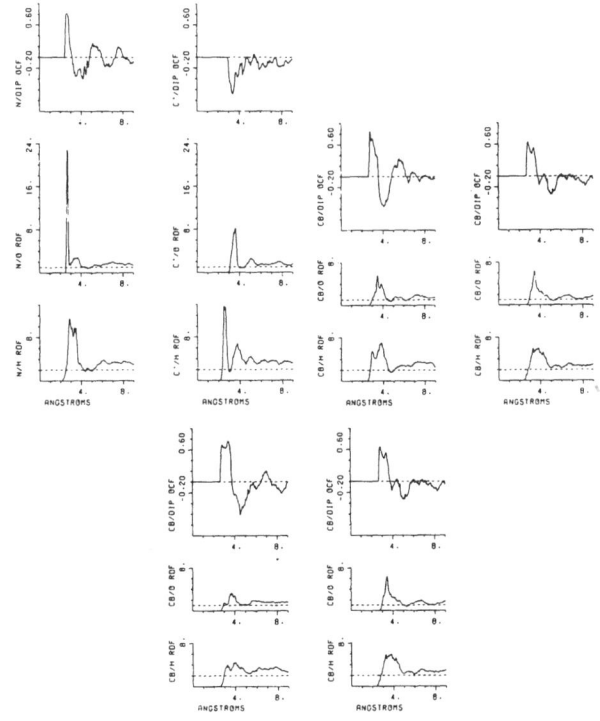

Figure 33. RDFs and OCFs for the N and C' regions of S3 (six inserts, top left) RDFs and OCFs for the C_B regions of S1 and S2 (six inserts, top right) and RDFs and OCFs for the C_B region of S3 and S2 (six inserts, bottom).

The N/H RDF for S1 (Figure 32, left inserts) peaks at 3 and 4.5 A; the corresponding N/O RDF exhibits a peak at 3 A and some smaller and broader ones between 4 and 6 A; the N/DIP OCF exhibits significant orientational order up to 4.5 A. In comparison with neutral glycine(161), one can notice a greater positional order in this region. Moreover, water molecules closest to the nitrogen atom of glycine tend to orient with their hydrogen atoms pointing toward it, whereas their serine counterparts have hydrogen atoms pointing away from it; this is likely to be due to the alcoholic hydroxy group. As for S2 (Figure 32, right insert), the N/H RDF exhibits two well resolved peaks at 3 and 3.5 A, the N/O RDF has a sharp peak at 2.8 A, and the N/DIP OCF exhibits a high orientational order in this region; the situation is qualitatively similar for S3 (Figure 33 (top left insert)); where the N/H RDF has a structured peak between 3 and 3.5 A, the N/O RDF has a sharp peak at 2.7A and a smaller and broader one at 3.5 A; these two peaks correspond to water molecules with opposite orientational orders. In comparison with glycine zwitterion, the large broad peak between 5 and 6A has moved inwards to 3.5A, and nearly disappears in the crystallographic conformer.

The C'/H RDF for S1 (Figure 32, left insert) exhibits a broad peak at 3.5 A, and the C'/O RDF has a main peak at 3.2 A and a broader, smaller one between 3.5 and 5 A; the region out to 5 A shows a significant orientational order. The C'/H RDF for S2 (Figure 32, right inserts) peaks at 2.7 and 3.7 A, the C'/O RDF has a peak at 3.5 A and a smaller and broader one at 5 A; there appears to be a significant orientational order up to 4 A; the results for the other conformer (Figure 33, top left inserts) are qualitatively the same. The general picture is qualitatively similar to the glycine one.

Finally, we compare the C_B regions (Figures 33 (top right and bottom inserts)). The C_B/H RDF for S1 peaks at 2.8 and 4 A, and C_B/O RDF has two peaks at 3.3 and 3.8 A; this solvation shell possesses a significant orientational order. C_B/H RDF for S2 has a broad peak at 3.5 A, and C_B/O RDF has also a peak at 3.5 A, and this solvation region exhibits a significant orientational order. The situation is qualitatively similar for S3, but with lower peaks in the RDF's. On the whole, this solvation is characterized by a higher positional order in the neutral molecule than in the non-crystallographic conformer of the zwitterion. This is not surprising if one assumes the C_B solvation to be essentially controlled by the neighbouring polar groups.

Let us now continue the structural analyses, using probability density maps. We calculated(165) orientationally optimized iso-energy contour maps for the solute-water interaction potential, in the ten planes z = -4, -3, -2, -1, 0, +1, +2, +3, +4, +5 A (and also z = +6, +7 A for S1) and also calculated both hydrogen and oxygen atom probability density (PD) maps for the corresponding regions: more precisely, for each z value, all solvent atoms with a z coordinate such that $|z - z| \leq 0.5$ A were accounted for in the appropriate PD histogram. We give here (Figures 34-36) hydrogen PD maps and oxygen PD maps superimposed to the corresponding iso-energy contour maps. The contour-to-contour separation in the PD maps is 0.2 atom/A^3, and 12.5 kJ.mol^{-1} for the iso-energy maps; the zero-energy and positive-energy contours are not shown. The distance between two consecutive tick marks in the perimeter corresponds to 2 A. For the sake of comparison, a similar analysis was carried out on glycine zwitterion. A number of water molecules, both belonging to immediate solvation regions and to farther ones were identified in the PD maps; a few general features, which are common to all the solutes we examined, can be pointed out.

1. Some water molecules directly solvate the amino or carboxy group.

2. There exist more or less rich and complex networks, consisting of hydrogen-bonded (hereinafter abbreviated H-bonded) water molecules, some which are situated in a close neighbourhood of the solute and/or are H-bonded to water molecules which, in turn, directly solvate the hydrophilic amino or carboxy group.

3. In some cases a "solvation gap" surrounds hydrophobic parts of the solute.

4. As for serines (Figures 34-36), there appears to be no evidence of any strong and direct H-bond to the alcoholic -OH group, owing both to the hydrophobic CH_2 and to the competition by the more powerful hydrophilic groups.

5. The previous points suggest that the detailed solvation structure is governed by a subtle balance between solute-solvent and solvent-solvent interactions. Indeed H-bond formation may stabilize solvent molecules in regions which are not particularly favoured by the solute-solvent potential, and, conversely, in a close neighbourhood of a COO^- group, the solute-solvent interaction may force two or more water molecules into a mutual organization far from the optimal one. Incidentally, it should be pointed out that, when we talk of H-bonds, we do not imply they are in their optimal geometry. On the other hand, the kind of structural information one can obtain from RDFs (where much detail has been averaged out) does broadly agree with qualitative predictions one can formulate on the basis of iso-energy contour maps.

Here we start to consider in detail the PD maps for S1 (see Figure 34). PD maps were calculated, both for water hydrogen and oxygen atoms around neutral serine, in the planes z = -4, -3, ..., 6, 7 A. Water molecule 9, hereinafter referred to as W(9) for simplicity, is H-bonded to the carboxylic -OH group; both its oxygen coordinates and orientation are essentially the optimal ones; W(2) is H-bonded to W(9) from below and W(13) from above; finally W(21) and W(26) extend the network further up. W(1) and W(3) directly solvate the amino group, and W(4) is possibly H-bonded to W(3). Pattern A is consistent with one water molecule outside the first solvation shell of the amino group, and H-bonded both to W(3) and W(4); pattern B is consistent with another water molecule H-bonded to W(1). W(5) is H-bonded to W(2), whereas W(6), W(7) and W(8) would appear to be H-bonded among themselves, and also, possibly to W(1) and W(9); W(7) is probably H-bonded to the C=O group. W(18), W(21) and W(22) are H-bonded among themselves and to W(19). Pattern M is consistent with a comparatively mobile water molecule, weakly bonded to its neighbours W(19), W(22) and W(26). W(25), W(27) and W(28) form another "chain" in the solvent network with W(25) bound to the solute. Pattern L is consistent with a comparatively mobile H-bonded chain, consisting of several molecules outside any first solvation shell. Other water molecules, such as W(24), would appear to belong to a "second solvation" shell. Apart - possibly - from W(25), there is no evidence of direct H-bond to the alcoholic -OH group. The structure-promoting effect of the carboxylic -OH group parallels and reflects the energy results.

Let us now consider in detail the PD maps for S2 (see Figure 35). W(1), W(2) and pattern A are H-bonded among themselves and belong to an outer solvation region. W(3) and W(4) are H-bonded to each other and also possibly to W(2), W(5) and W(6). So to speak, they complete the first solvation shell below the molecular plane. W(5), W(9) and W(17) are H-bonded to the protons of the NH_3^+ group; W(10), W(11), W(12) and W(13) have their oxygen atoms essentially localized in the region $0.5 \leq z \leq 1.5$ A; they appear to be H-bonded among themselves, and also, probably to W(9) and over W(17). W(7) solvates O2 below the molecular plane, and W(8) is H-bonded to W(7); W(6) solvates O1 below the molecular plane.

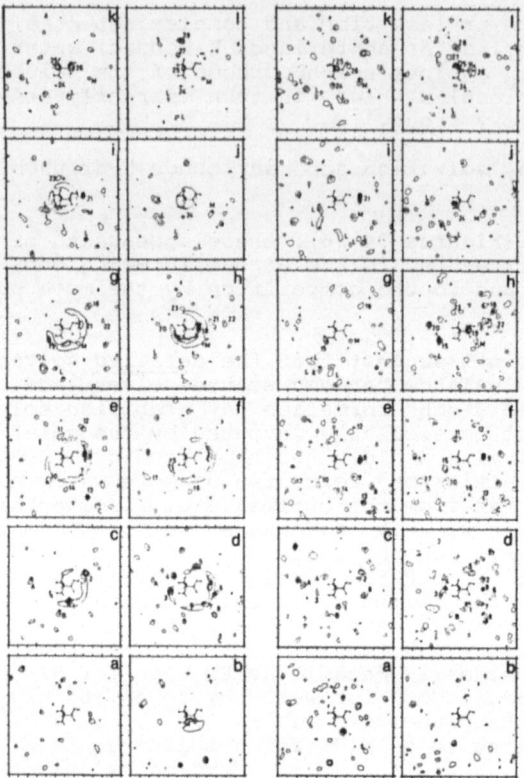

Figure 34. <u>Left Inserts</u>: Oxygen - atom PD and iso-energy contour maps for S1, for the planes:

a) z = -4 A, b) z = -3 A, c) z = -2 A, d) z = -1 A,
e) z = 0 A,   f) z = 1 A,   g) z = 2 A,   h) z = 3 A,
i) z = 4 A,   j) z = 5 A,   k) z = 6 A,   l) z = 7 A.

<u>Right Inserts</u>: Hydrogen - atom PD maps for S1; same meaning of the letters in the left inserts.

W(14), W(15) and W(16) have their oxygen atoms localized in the region $0.5 \leq Z \leq 1.5$ A, and probably form a somewhat distorted H-bonded network (compare with the corresponding energy results); W(14) and W(16) solvate the O2 from above the molecular plane. Pattern X is consistent with three H-bonded water molecules, one of which solvates O1 from above the molecular plane. W(18) is probably H-bonded to W(10) and extends the previously mentioned network. W(19) and W(20) are H-bonded and outside any first solvation shell; pattern Y is also consistent with another H-bonded chain in a second solvation shell. Pattern Z suggests two H-bonded solvent molecules on top of the solute.

Let us now pass to a detailed analysis of the PD maps for S3 (see Figure 36). Patterns L, M and P are consistent with H-bonded chains outside any first solvation shell; a similar conclusion can be drawn for pattern Q, which however is not far from the alcoholic -OH group. W(8), W(20) and W(26) are H-bonded to the protons of the NH_3^+ group. W(9) and W(12)

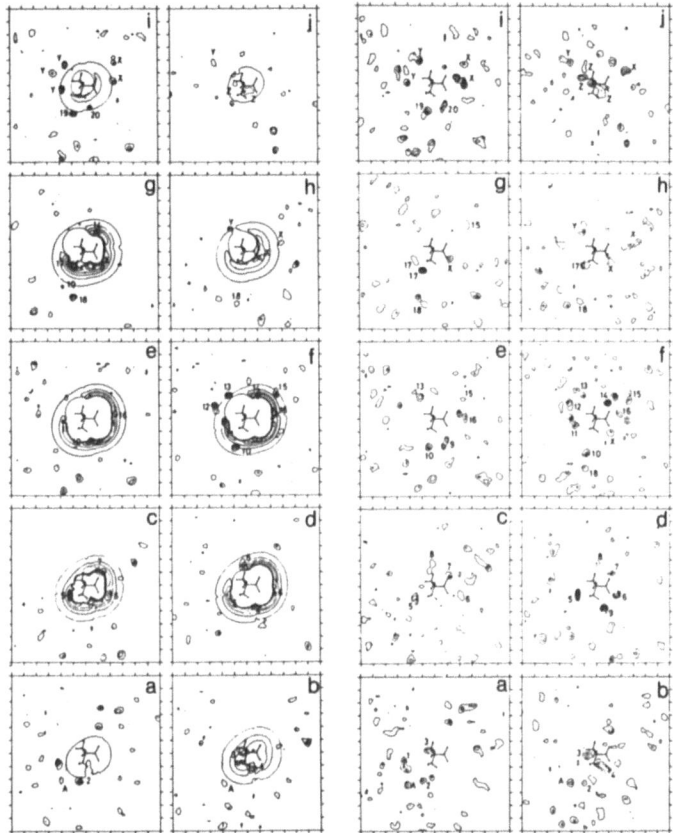

Figure 35. Left Inserts: Oxygen - atom PD and iso-energy contour maps for S2, for the planes:

a) z = -4 A, b) z = -3 A, c) z = -2 A, d) z = -1 A,
e) z = 0 A, f) z = 1 A, g) z = 2 A, h) z = 3 A,
i) z = 4 A, j) z = 5 A.

Right Inserts: Hydrogen - atom PD maps for S2; same meaning of the letters as given at left inserts.

directly solvate O1 and O2 respectively below the molecular plane. W(28) and W(29) similarly solvate the carboxy anion above the molecular plane. W(2), W(3), W(4) and W(5) form an H-bonded chain, with W(5) probably H-bonded to W(8); W(1) and W(6) are H-bonded, and W(1) would appear to be H-bonded to W(9); W(19) and W(23) are H-bonded to W(20). W(13), W(14), W(15), W(16), W(17, W(18), W(19) and W(21) form a rather complex H-bonded network which connects the C' and the N solvation regions without any noticeable gap. W(32) is probably H-bonded both to W(26) and W(28). W(30) and W(31) would appear to form a weak double bridge between W(28) and W(29). It is interesting to compare the energetic and structural results for the two conformers.

S2 shows a stronger solvation around both ionic ends, that is, both a

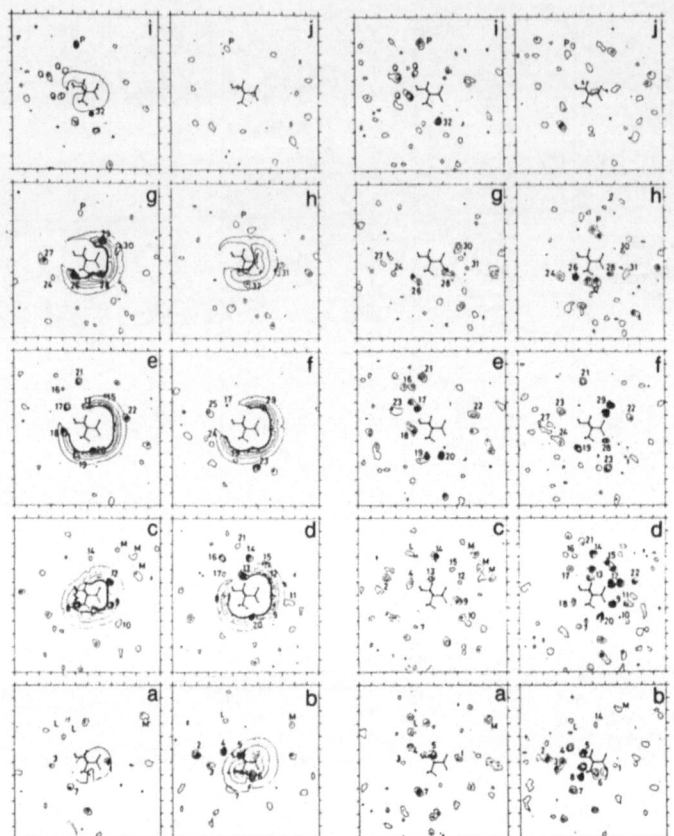

Figure 36. <u>Left Inserts</u>. Oxygen - atom PD and iso-energy contour maps for S3; same meaning of letters as in Figure 35.

<u>Right Inserts</u>. Hydrogen - atom PD maps for S3; same meaning of letters as in Figure 35.

stronger and/or longer ranged solute-solvent interaction, and (especially for COO$^-$) a more profound disruption of the mutual organization of solvent molecules. This reflects the fact both protons in the CH$_2$ group point "outwards", and produce a more pronounced "solvation gap".

On the other hand, in the S3 conformer, the CH$_2$ group interferes to a greater extent with both ionic ends, thus reducing their "solvation" (in the same sense as above), especially for the anionic end. In this conformer both a hydrophilic -OH and a hydrophobic C-H bond point "outwards", and their effects will partly balance. These facts account for the reduced size of the "solvation gap" and the existence of a more rich and complex H-bonded network in the second solvation shell.

To conclude, we shall analyze in detail the PD maps for glycine zwitterion (see Figure 37). W(1) and W(2) are H-bonded and also possibly H-bonded to W(4) and/or W(9). W(4), W(9), W(11), W(14), W(18) and W(19) solvate the NH$_3^+$ group; W(10) and possibly W(19) belong to the second

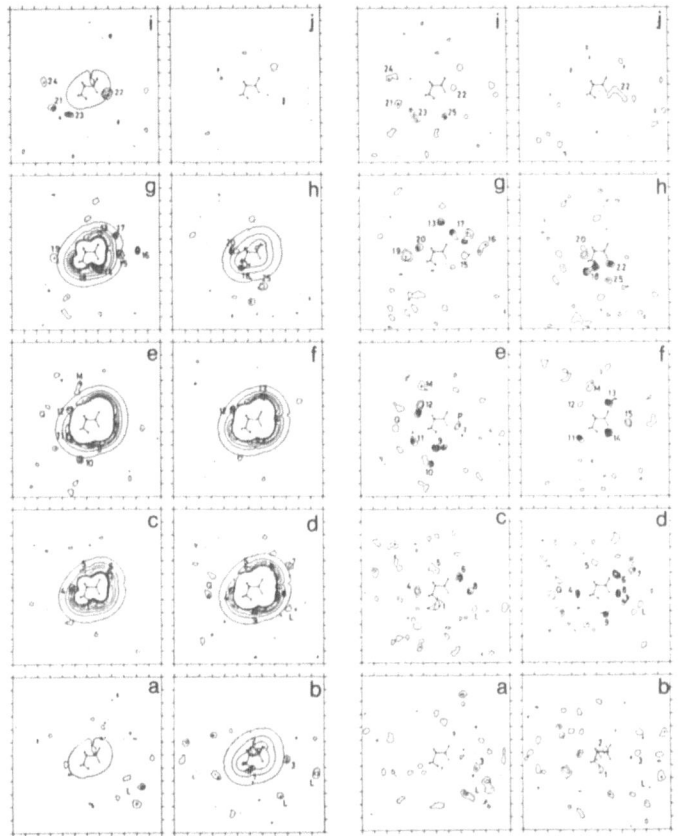

Figure 37. Left Inserts. Oxygen - atom PD and iso-energy contour maps for glycine zwitterion; same meaning of letters as in Figure 35.

Right Inserts. Hydrogen - atom PD maps for glycine zwitterion; same meaning of letters as in Figure 35.

solvation shell; there are also H-bonds among these molecules. W(6) and W(8) solvate the COO$^-$ anion from below; W(13), W(14), W(15) and W(17) solvate it from above the molecular plane. W(17) is H-bonded to W(6), W(16) is H-bonded to W(15) and W(17), and W(22) is H-bonded to W(15). W(25) is likely to be H-bonded both to W(18) and W(22). Pattern P is consistent with one water molecule on the molecular plane, weakly bound to the carboxy anion. W(18), W(19), W(20), W(21), W(23), W(24) would appear to build up an H-bonded chain which starts in the first solvation shell of NH$_3^+$; W(4), W(5), W(12), pattern M and W(13) form some kind of (presumably weakly H-bonded) bridge (essentially situated on the molecular plane) joining the N and C' solvation regions; W(3) and pattern L are outside the first C' solvation shell. It is interesting to note that, although the solute molecule possesses a C_s symmetry, the hydrogen-bond interaction and the resulting solvation structure do not.

Comparison between glycine and serine shows that the presence of the alcoholic group enhances the solvation of the neutral molecule, whereas it has a more subtle effect on the zwitterion.

Thus "hydrogen bonded" filaments of water, often crosslinked among themselves appear not only in ionic solution (see Section 5.2), but also in simple organic molecules, like the zwitterions of the amino acids. Below, the same information will be presented in the study of the water structure of nucleic acids. These filaments are most likely the key aspect for fast, long distance deprotonation at one side and protonation at another site in living organized matter.

5.8 ENZYME-WATER INTERACTION IN SOLUTION: A PRELIMINARY STUDY ON LYSOZYME

The structure and function of proteins strongly depend on their electronic properties and on their interactions with the solvent. Therefore, more information concerning these properties may add significantly to our understanding of such systems. Indeed, the reaction between an enzyme and its substrate is an event affected not only by the geometry of the two interacting species, but also by the electronic environment at the active site. It is well established that enzyme activity depends on the pH and that at each pH a certain fraction of the amino acids is ionized. These tend to regain electro-neutrality either by forming internal salt bridges or by attracting counter-ions from the medium (ions are present even in the crystal form of a protein). Such an event usually results in a re-orientation of the water molecules about the charged regions. Thus, the total electrical field is due to the residual charges between the neutral residues, to the charges that are partially neutralized by counter ions, and to those of the polar side-chains without counter ions.

As known, it is difficult to model the solvent molecules, and hence to study their interactions with the protein. Such solvent molecules are present even within protein crystals but, because of lack of order or crystallinity limitations, often their coordinates are not reported in X-ray studies. However, in the few cases where solvent coordinates are available, they could serve as a starting point for relating to solution properties.

Even the most elaborate list of water coordinates obtained by X-ray crystallography, is presently deficient since it covers only a fraction of the bound water. Moreover, even for this partial list there is an uncertainty in their locations, and their sites are not fully occupied. In fact, since some of the long, charged, side-chains, which are on the surface of the protein, are relatively free to move, there is also uncertainty in their coordinates.

Although the information obtained from X-ray studies is incomplete, it is the only unambiguous structural observation, and eventually one has to compare any computational results with the experiment.

Some of the main problems concerned with the interactions between water and proteins are: (a) how can one define the "solvation shell" and is this shell made of a mono-layer or must water molecules that are not in direct contact with the protein be considered as part of outer shells? (b) to what extent is this shell ordered, or what is the life-time of any water molecule in a fixed location? (c) What is the fraction of the solvent in the crystal which is chemically bound to the protein, and to which extent is the rest of the water free to move? (d) Are all the potential water ligands on the surface of the protein (that is, hydrophilic side-chains and amino and carbonyl groups of main chains) really bound to water? (e) What is the correlation between the crystallographic results and the structure in solution? We shall attempt some limited and preliminary answers to such questions.

The computational problems related to studies of proteins in solution present at least two types of difficulties. The first was the construction of a library on interaction potentials (ab initio) to describe the intermolecular interactions between the three-atom water and the many-atom enzyme. This library is now available.

A second difficulty was encountered when one attempts to use our atom-pair potentials (or, in general, any potential) to perform a Monte Carlo simulation of the structure of the solvent that surrounds an enzyme. Unless one restricts oneself to study only the molecules of water that are present in the enzyme in a crystalline form (a few hundred molecules and contained in well-defined regions of the unit cell), the number of water molecules to be considered in order to correctly represent the first few solvation shells is of the order of several thousands, even for relatively small enzyme such as lysozyme.

In the following sections we present the preliminary results obtained for lysozyme(166); this enzyme was chosen because of the availability of literature data and unpublished data on solvent positions(167).

The protein is placed in a rectangular box of dimensions larger than the protein. The box is cross-sectioned with a number of parallel planes (taken with the same plane-to-plane distance), each plane is subdivided into an orthogonal grid. For each plane, the oxygen atom of a molecule of water is placed on the grid point and the hydrogen atoms are rotated until one finds the position of energy minimum. The computed energies of interaction of a water molecule with a protein are then displayed as isoenergy contour maps. A complete description of the interaction energies around the surface of the protein may be obtained by stacking all the planes in sequence. In order to obtain a preliminary and rough list of potential water locations, we determined the deepest point in each energy hole and located one water molecule there. If this molecule did not cover all the space available in the particular hole, we located more water molecules in points that are at least 2.8 A apart from the existing ones.

To simulate the solvent-protein interaction the protein structure is assumed to be the one indicated by X-ray data. Such data, however, do not show the hydrogen coordinates; as a consequence, the presence of ionic residues cannot be unambiguously detected (indirect determination is obtained from the presence of counter-ions). On the other hand, in any simulation of the interaction of a molecule of water, the orientation is a most important factor; it follows that the hydrogen atoms must be included in the water molecule as well as in the protein.

On the outset of a simulation, a decision must be taken on the model to be simulated. Essentially, there are four alternative models to simulate the interaction with water: (1) a neutral protein in solution, (2) a neutral protein with the water present in the crystal, (3) a charged protein and counter-ions in solution and (4) charged protein and counter-ions in the crystal. We have selected the first model assuming that the conformation in solution remains essentially unchanged relative to the conformation in the crystal. Concerning the method, we notice that the interaction of a protein with the solvent could be, in principle, evaluated by the standard Monte Carlo approach.

Pilot studies have indicated the need of a rather large number of molecules of water, even for small solvated systems.

Thus, it seems clear that to properly simulate an enzyme in solution, the inclusion of few thousands of water molecules is required. An independent evaluation(168) has set at about 500 the number of molecules of water that can be explicitly simulated in a Monte Carlo computation; this limitation was essentially set by the power of currently available computers. This limit is not a hindrance, if one wants to study water occluded in the protein crystal, but is a bottleneck if one wishes to simulate an enzyme in water solution. New programming techniques and hardware (vector oriented) are, however, abailable; with these tools, 2,000 to 4,000 water molecules can be simulated.

Another observation concerns the possibility of including in the standard Monte Carlo only few molecules of water, namely those molecules that constitute the first solvation shell. This presents a problem, however, as indicated in a study on the interaction of phenylalanine (Phe) with water (169). The energy maps have been used to determine the energy minima of the potential surface and in the absolute minimum a molecule of water was placed with the usual technique. Then the potential surface was recomputed for the system Phe + (H_2O) interacting with a second molecule of water and the second molecule of water was placed at the new absolute minimum. This process was repeated 24 times, yielding the Phe + 25 (H_2O) cluster. The most interesting result obtained in this way is that the molecules of water do not surround uniformly Phe, but they start by clustering around the $-NH_2$ and -COOH groups and only after that, they surround hydrophobic portions of Phe. The minima around the aromatic ring are generated by the water-water hydrogen bonds rather than by water-Phe interaction. This is the behaviour which one expects on the basis of hydrophilic and hydrophobic interactions and it questions the accepted meaning of "first solvation shell" as a complete shell around a molecule in solution.

If for first solvation shell one merely states the fact that when a protein is in solution, a set of molecules of water are found and are nearer to it than the remaining ones, then one has a unique definition. However, some of the molecules can be strongly bound to the protein, some loosely bound and some not bound at all and present only because of the water-water interaction.

We explored a new technique that has been empirically developed during this study of lysozyme and that is here exposed in a somewhat formalized manner. The method has an application range limited to systems where one can construct a reasonable starting distribution that requires only a slight perturbation in order to yield a reasonable final description of the system. In other words, we assume that a relatively good "zero order" description is available and therefore a "first order" correction is sufficient. It is noted that we compare our computational results with diffraction studies of given resolution that sets the upper limit of the resolution meaningful to look for in our simulation. We describe our method by a sequence of steps that correspond to different phases of the computations.

In the <u>first step</u> the protein, as available form X-rays data and with addition of the hydrogen atoms, is enclosed in a rectangular box that exceeds the dimensions of the protein. The rectangular box is divided into a three dimensional grid where the distance between two successive grid points is less than the water-water distance, that is 2.8 A. At each grid point the oxygen atom of a molecule of water is placed and the hydrogen atoms orientations are optimized by performing a number of rotations. It is noted that step one corresponds to the above reported method for studying the interaction of a protein with a single molecule of water. We designate with M_1 the number of molecules of water considered in the first step.

In the <u>second step</u>, one retains only those molecules of water located in regions where the interaction with the protein has a minimum or, at least is attractive; in this way one obtains a set M_2 of water molecules with $M_2 < M_1$. Clearly the M_2 molecules could have been chosen with different criteria, depending on the type of information one wishes to obtain. For example, one could constrain the choice to water molecules located within a given range of distances from the protein.

In the third step, the water-water interaction is included and each water is rotated to optimize the interaction energy both with the protein and the water molecules. We obtain a set M_3 with $M_2 = M_3$ but where the orientation of the waters in M_2 differs from the orientation of those in M_3. In this way, the water-water interaction is considered as the perturbation on zero-order distribution.

In the fourth step, the M_3 molecules are translated along the x, y, z directions by a fixed increment, that is a fraction of the grid interval (of the order of 0.2-0.5 A). This limited translation is intended to allow each molecule of water to reach its proper minimum position when in the presence of other molecules of water. The six translations and the original position create a set of 7 x M_3 molecules of water; from this set a final set is called M_4 and is equal to M_3. Finally, the M_4 molecules are again allowed to rotate around their oxygen atoms. The computer time saved using the outlined method, compared with the computer time used in a Monte Carlo simulation for a comparable system, is quite significant. However, the results obtained are less reliable.

As described above, the box of the protein was sliced into sections parallel to the Z = 0 plane with intervals of 2 A. Each plane is divided into a grid of 2 A. The protein was considered to be neutral; the charges of the backbone are different from those computed for the individual amino acids (from 5 to 20%) since the backbone charges are those obtained from formyltriglycilamide(82).

There are some "holes" inside the core of the lysozyme molecule; such holes have been detected in the computation and we found six possible positions where one molecule of water could be placed. The coordinates for the oxygen atom and the binding energy for the six molecules of water are given in Table 38. A comparison with the computed coordinates and those found by X-ray analysis by Imoto(170) and by Moult and Yonath(171) is also present in the table.

Particular attention was given to the active site of the enzyme. In Table 39 we list the coordinates of the oxygen atom for the water molecules that are located in close proximity of the residues that limit the "binding pocket".

Table 38. Computed Locations For Six H_2O Molecules
In The Interior of the Lysozyme Molecule(a)

x	y	z	Energy (kcal/mole)	Interacting atoms(b)	From Residue(s)	Note
2.0	17.0	19.0	-8.3	O,HN1,OG	58,56,91	(c)
0.4	12.1	17.5	-4.2	N	88	(d)
-3.9	10.4	4.8	-7.4	-	-	(e)
0.0	13.0	17.0	-5.7	HOG,N	91,88	(f)
4.0	22.0	25.0	-5.7	HN1	57	(f)
14.0	15.0	27.0	-16.1	HOG,OG	60,69	(g)

Notes: (a) Distances are in A.
(b) Up to a distance of 3 A.
(c) Reference 170: x = 1.4, y = 16.7, z = 19.9;
    Reference 171: x = 3.10, y = 17.0, z = 19.2.
(d) Reference 170: x = -0.6, y = 11.1, z = 19.5.
(e) Reference 170: x = -1.9, y = 12.4, z = 4.8
(f) Not found by X-ray diffraction studies.
(g) Reference 171: x = 15, y = 15, z = 27.

Table 39. Water Molecules Located in Energy Minima In The Region of The Binding Pocket

Energy (kcal/mole)	Oxygen coordinates, Å			Atoms at distances up to 3 Å	From residue(s)
	x	y	z		
-13.90	8.0	34.0	20.0	HND1, ND2; HG	113; 119
-10.24	8.0	31.0	17.0	HG1; HND2, ND1; HA	109; 113; 110
-7.74	10.0	33.0	16.0	HND2;	113
-7.56	9.0	27.0	19.0	CG1, HG2; OE1 (3.3 Å)	109; 35
-7.46	18.0	23.0	21.0	O; HA	45; 46
-7.39	11.0	23.0	16.0	CG, ND2; HG11, HND1	44; 25
-7.68	17.0	21.9	19.0	HN1, O, N	45
-5.77	14.0	23.0	20.0	HB2; ND2	44; 46
-4.97	6.0	26.0	26.0	HA; HN1	108; 109
-.5.09	3.0	22.0	22.0	HOD2; HND1; ND2 (3.4 Å)	46; 52

We compare our list with a preliminary, unrefined list of peaks that were found in the different electron-density maps of triclinic lysozyme and to which solvent molecules with occupancies higher than 0.5 were assigned(167). From this comparison it was found that there are at least 10 water molecules that have almost the same locations in the two lists and lie in fairly deep energy "holes". A total list that accounts for about 30% of the experimentally observed water molecules is given in Table 40. It is of interest to mention that 4 of these locations were assigned as NO_3^- ions from the X-ray results. In Table 41 we present the coordinates of the water molecules nearest to the observed one, according to Step 4.

In the following, we report the results obtained following our step-wise method. In the first step, the sides of the rectangular box are chosen to be from -22 to 26 Å for the x axis, from 1 to 44 Å for the y axis, and from -5 to 40 Å for the z axis. The initial set of M_1 molecules is equal to 12096. Most of such molecules correspond to grid points inside

Table 40. Coordinates Of The Water Molecules Nearest To The Observed Ones (At Step 2, See Text)

Distance	Oxygen Coordinates, Å (a)			Energy (kcal/mole)
	x	y	z	
0.44	-15.00	16.00	1.00	-7.11
2.34	-17.00	18.00	1.00	-5.55
2.09	4.00	23.00	5.00	-5.01
2.38	-2.00	33.00	7.00	-8.74
2.13	9.00	12.00	9.00	-5.82
0.97	-19.00	22.00	9.00	-8.27
2.44	7.00	26.00	9.00	-7.36
1.92	0.00	5.00	11.00	-6.34
1.60	-15.00	16.00	9.00	-5.50
1.30	-17.00	26.00	9.00	-6.23
1.88	6.00	35.00	11.00 (b)	-5.93
1.22	-3.00	34.00	9.00	-19.93
2.17	9.00	18.00	9.00	-5.15
1.53	-12.00	29.00	11.00	-8.32
1.03	-14.00	25.00	11.00	-6.90
2.37	2.00	5.00	11.00	-6.88
1.53	8.00	35.00	13.00	-5.50
2.33	-8.00	9.00	15.00	-7.95
1.62	-14.00	33.00	15.00	-5.82
1.19	8.00	31.00	17.00 (b)	-10.24
2.08	-6.00	35.00	15.00	-7.61
1.47	-8.00	7.00	19.00	-11.34
2.42	8.00	27.00	17.00	-5.79
1.12	2.00	17.00	19.00	-8.34
1.56	8.00	35.00	21.00	-7.66
2.06	8.00	35.00	23.00	-6.09
1.32	-2.00	37.00	25.00	-8.77
1.00	14.00	15.00	27.00	-16.10
2.37	-9.00	18.00	29.00 (b)	-5.75
1.28	-13.00	26.00	33.00	-5.85
1.52	16.00	19.00	35.00	-5.02
1.93	5.00	6.00	33.00	-5.81

Notes: (a) Computed at step two.
(b) It refers, likely, to the position of an ion.

Table 41. Coordinates Of The Water Molecules Closest to The Observed Ones (At Step 4, See Text)

Dist. (a) (Å)	Oxygen Coordinates, Å			Energy (kcal/mole)
	x	y	z (b)	
0.40	-15.00	15.70	1.00	-1.40
2.20	-19.00	20.00	1.30	-9.30
2.30	4.00	31.00	5.30	-9.50
1.80	6.00	19.00	5.30	-6.10
2.30	9.00	12.00	9.30	-10.80
2.00	-19.00	20.00	9.00	-6.90
2.40	7.00	26.00	9.30	-9.10
2.40	1.00	6.00	9.30	-14.50
1.50	-15.00	16.00	9.30	-9.00
2.20	-19.00	24.00	9.00	-4.00
1.60	6.00	35.00	10.80 (c)	-7.30
1.20	-3.00	34.00	9.00	-11.30
2.20	8.70	18.00	9.00	-6.50
1.90	-15.70	25.00	11.00	-18.00
2.10	-6.00	9.00	13.30	-16.30
1.70	-8.00	11.00	13.30	-21.50
0.50	-18.00	25.00	13.30	-5.80
0.70	-20.00	21.00	13.30	-5.00
2.20	8.00	33.00	13.00	-2.50
1.10	8.00	31.00	16.70 (c)	-5.40
1.80	-6.00	35.00	15.30	-13.50
1.40	-8.00	7.00	19.00	-16.10
0.50	10.00	27.00	18.70	-6.50
1.80	-18.00	25.00	17.30	-4.70
1.10	2.00	16.70	19.00 (c)	-8.60
1.00	-13.70	33.00	19.00	-10.10
2.40	-2.00	7.00	19.30	-11.60
1.70	8.00	35.00	21.30	-8.60
0.80	8.00	23.00	21.00	-6.10
1.50	5.00	24.00	22.30	-11.80
1.70	-11.00	12.00	22.30	-13.00
1.30	-2.00	37.00	25.00	-6.60
1.10	4.00	22.70	25.00	-4.00
1.80	12.00	9.00	25.30 (c)	-4.50
1.00	14.00	15.00	27.00	-15.00
1.40	-10.00	15.00	27.30	-5.80
0.80	4.30	29.00	27.00 (c)	-5.60
1.00	8.00	31.00	37.00	-5.40
1.30	9.00	6.00	28.70	-7.90
2.00	-11.00	26.00	33.00	-7.10
1.80	5.00	5.70	33.00	-9.70
1.30	-5.70	19.00	35.00	-7.20

Notes: (a) From the nearest observed water molecules.
(b) Computed.
(c) Corresponds likely to the position of an ion.

the lysozyme core and are therefore discarded, with exception of the six previously mentioned molecules (Table 38). By selecting those water molecules with attractive interaction energy with the protein, a set of about 1600 molecules was retained. The previous discussion concerning the interaction with only one molecule of water deals with this set. By using not only the protein-water interaction energy but also the water-water distance as selection criterion the set is again reduced and we obtain the M_2 set with $M_2 = 307$.

Figure 38 presents 9 sections with the isoenergy contour maps. Figure 40 shows the 9 sections of Figure 38 with the identification of the position and the orientation of those molecules (out of the 307) that happen to have the oxygen in the x, y plane or displaced by ±0.3 A along the z axis. From these figures one may conclude that the protein consists of regions of different level of hydrophobicity. In particular, water in the region of residues 97-100 (see inserts h) which are the components of the flexible loop (172) are very hydrophilic. In Figure 39, we present a three-dimensional view of the nine sections obtained by considering the hard-core iso-energy. In Figure 41 we report the results of our analysis obtained for the set of 307 molecules at step. 4.

In Figure 41, insert A, we present the water-water interaction energy; the top insert a describes the distribution (frequency) of the 307 molecules of water as obtained at step two. The molecules, being oriented so as to optimize only the water-protein interaction, show - as expected

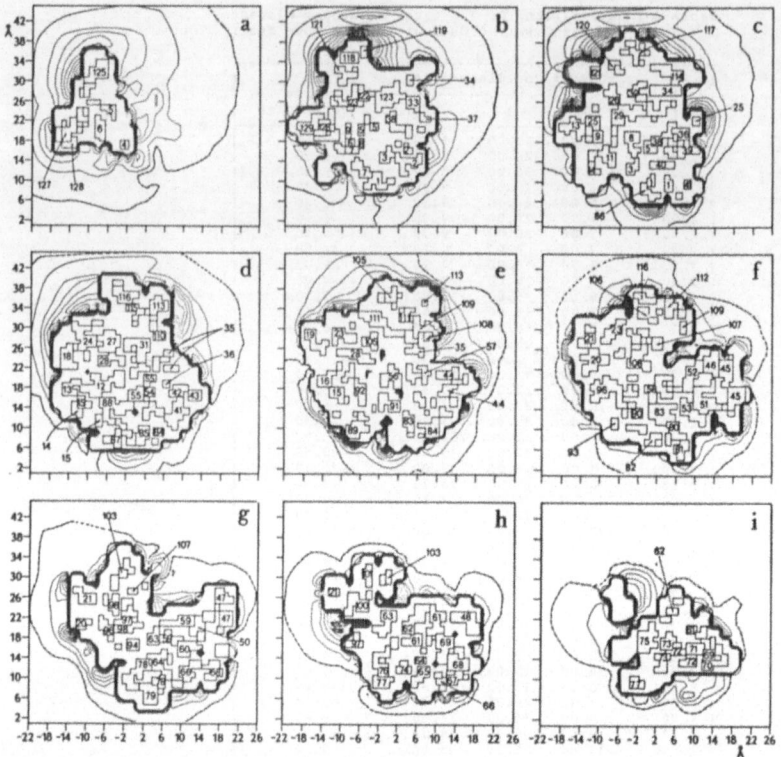

Figure 38. The nine inserts correspond to the iso-energy contour maps obtained by computing a single water molecule interacting with lysozyme at nine cross sections. The z values are from 1 to 35 A in steps of 4 A. See Figure 38.

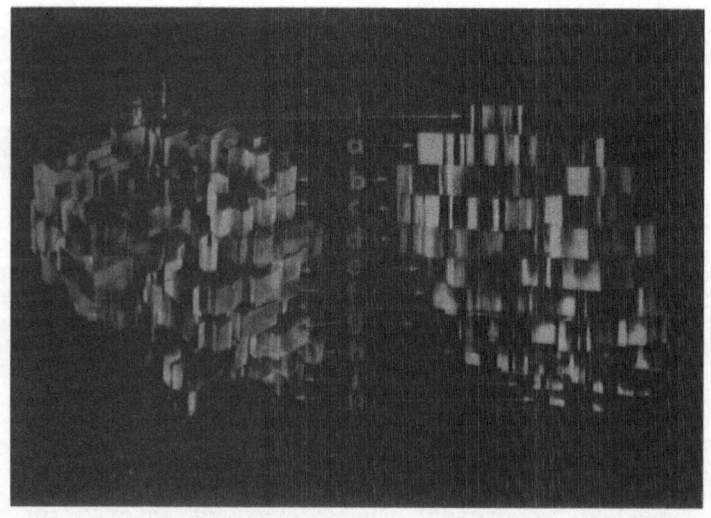

Figure 39. Model of Lysozyme (at two different orientations) corresponding to the hard-core iso-energy maps.

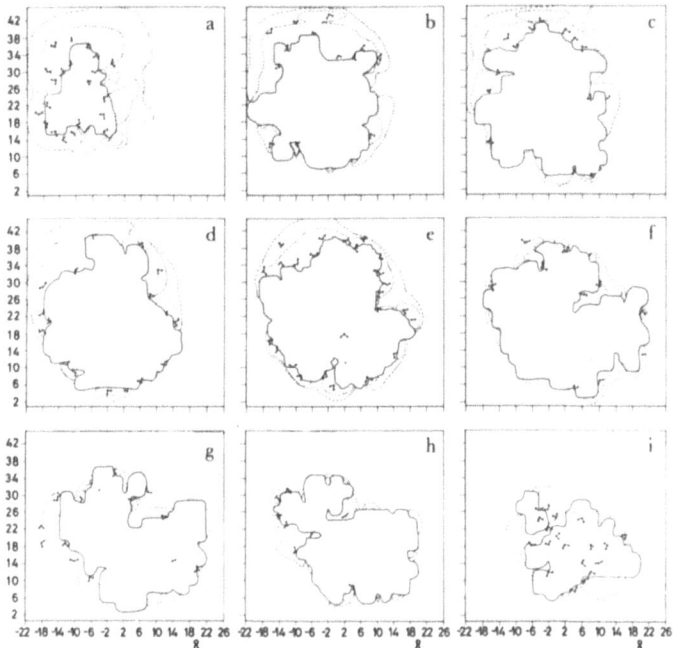

Figure 40. The nine inserts represent molecules of water predicted to be at the indicated position and orientation, with the oxygen either on the plane or displaced by almost 0.3A. For inserts (a) and (i), a number of water molecules are added and correspond to farther away molecules. The dotted contours correspond to -1.0 and -02.0 kcal/mole. The solid line corresponds to the beginning of the repulsive region.

- no net water-water interaction: the distribution peak is relatively narrow and has its maximum at about -2 kcal/mole; few molecules have a water-water interaction as large as - 4 kcal/mole and those correspond likely to dimers or small clusters of water. The inclusion of water-water interaction in the optimization of the water molecules orientation leads to the situation displayed in the middle insert b of A. The distribution shows four peaks, one at about zero water-water interaction energy (molecules of water far from any other water molecule), one at -4 kcal/mole (likely dimers), one at -8 kcal/mole (likely trimers) and one at -10 kcal/mole (small clusters). This situation remains virtually unchanged, even after translation of the molecules and in the insert at the bottom of insert A we notice the five peaks at 1.0, -5.0, -8.0, -11.0 and -13.0 kcal/mole. The water-water perturbation acting on the originally nearly gaussian distribution of interactions (top insert) has brought about a net stabilization of the 307 water molecules, indicated by the net shift to higher attraction energies.

In Figure 41 (insert B) we analyze the water-protein interaction energy. The distribution at step two (top insert) is rather symmetrical with a peak at -4 kcal/mole and the asymmetry of the distribution is in the sense to favour the water-protein interaction. After having "turned on" the water-water interaction and let the molecules undergo rotations to optimize the orientation, the water molecules are less bound than before (see b of insert B) and the above-noted asymmetry is much more pronounced. This situation remains nearly unmodified to the end of the simulation (insert c of B).

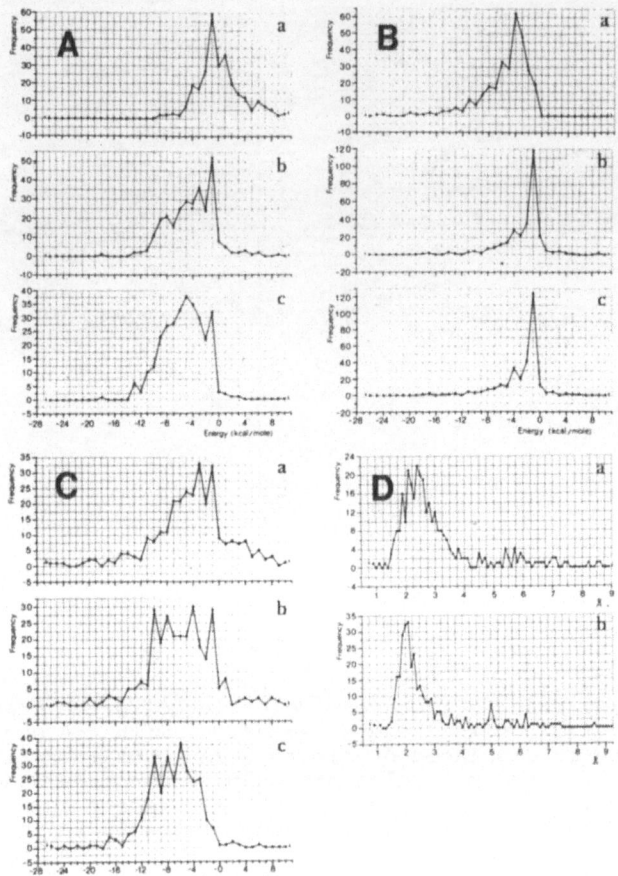

Figure 41. Energy distribution of water molecules interacting with lysozyme. Insert A. The inserts (a), (b) and (c) represent the water-water interaction energy computed at steps 1, 3 and 4, respectively. Insert B. The inserts (a), (b) and (c) display the water-enzyme interaction energy at steps 2, 3 and 4, respectively. Insert C. The inserts (a), (b) and (c) represent the total interaction energy (namely water-water plus enzyme-water) at steps 2, 3 and 4, respectively. Insert D. Distance distribution of water molecules interacting with lysozyme. The two inserts correspond to distances computed at steps 2 and 4.

The total interaction energy is the combined result of the water-water interaction and of the water-protein interaction. The somewhat symmetrical and narrow distribution with the peak at about -30.0 kcal/mole in Figure 41 (top of insert C) broadens considerably by "turning on" the water-water interaction (middle insert of C) and, after the translations, some structure becomes visible.

The structure could be an artefact of our simulation, but is very tempting to identify the peaks as due to single molecules of water, dimers, trimers and clusters of water interacting with the protein. Thus, one can state that, whereas a hydrogen bond in water is about -5 kcal/mole, a hydrogen bond between water molecules is weaker (perhaps half as much) when in the hydrophilic region of the proteins. This result is an expected: the protein-water attraction tends to orient the water molecule differently from the optimal orientation existing in liquid water or ice;

thus, the average hydrogen bond of water-water is smaller in the hydrophilic region than in water.

Let us consider now the distribution of our set of 307 molecules of water as function of the distance from the enzyme.

As the "distance" we mean here the minimum value computed for the distance between an atom in the protein (hydrogens included) and one of the three atoms of water. In the top insert of D (in Figure 41) the 307 molecules, obtained at step two, show a nearly symmetrical distribution with a maximum at about 2.5 A. Larger distances can be considered as referring to dimers or small clusters. We believe that the fine structure can be correlated with different hydrogen bonds N-H-O, O-H-O, H-O-H. At the end of the process (bottom insert of D) the distribution is shifted towards smaller distances (peak at 2.0 A) as a result of the "pressure" exerted by the perturbation of the solvent (water-water interaction). It is of interest to note that over fifty molecules of water are in the tail from 3.5 A to larger distances. The interpretation is likely to be related to our previous discussion on the meaning of the "first solvation shell".

In Table 41 we report the oxygen coordinates for those molecules of water where acceptable agreement with X-ray data(167) has been found. The list corresponds to about 50% of the total number of water molecules determined by X-ray analysis.

Before attempting to characterize the arrangement of the water molecules around lysozyme, it is important to recall the inherent weak points of the method. First, the coordinates provided for the protein were obtained by X-ray diffraction analysis of a crystalline system(171) and here we try to understand the protein-water interactions in solution. However, there is a high similarity between the coordinates in the triclinic and in the tetragonal form of lysozyme(167, 171).

Second, the X-ray diffraction data were collected to 2.5 A resolution, which limits the accuracy of the coordinates. However, the triclinic form diffracts well beyond 1 A, which results from high crystallinity; therefore the protein in this crystal form is very ordered.

Third, the experimental coordinates for the water molecules were obtained in the crystal state. However, different electron density-map analysis of the water arrangement in the two crystal structures mentioned above shows that the locations of most of the water molecules not involved in protein-protein contacts, are nearly the same. Hence, we assume that the structure of the tightly bound water molecules in the crystal is very similar to that in solution.

The fourth point is quite severe and concerns the electrostatic state of the protein molecule. At present, we assume a completely neutral state as an extreme case. This is not unjustified, since even if charges are present on the protein, there are probably counter-ions next to them. However, we are aware of the fact that in order to complete this study one has to look at the "fully charged" protein as the other extreme case. We note that, recently, a preliminary list of locations of counter-ions was provided from the crystallographic studies.

From our study we can conclude that the assumption of a single shell around the protein can be accepted only as a very crude zero-order approximation. The experiments on the Phe cluster, Monte Carlo computations and our results on lysozyme point out rather convincingly that there are energetically stable positions for a water molecule at a distance larger than it can be accounted for by the "one shell" hypotheses.

The data of Table 41, with the predicted coordinates of water, should be compared with those obtained in a Monte Carlo study of the water present in lysozyme crystal(168), but unfortunately the latter have not been published. The only statement available on this regard is that 49 water molecules are predicted to occupy the same environmental niches found in a list of about 80 molecules of water experimentally determined by X-ray diffraction. From a figure presented in the Monte Carlo simulation(168), one could tentatively deduce that the water probability maps have a rather low resolution, possibly because of the relatively small number of configurations considered (400,000 configurations were used to compute the distribution of water around the enzyme, namely less than needed in our studies of water around small molecules or a single ion). However, the energy data reported in Figure 40 can be compared with the equivalent Monte Carlo data and indicate a substantial agreement. This constitutes an interesting check of our method, even at this primitive stage.

Since this analysis was performed, we have improved our Monte Carlo program: inclusion of about 2×10^3 water molecules seems feasible even without vector oriented software and hardwares.

5.9 THE WATER STRUCTURE IN THE ACTIVE CLEFT OF HUMAN CARBONIC ANHYDRASE/B.

The formation of H_2CO_3 from H_2O and CO_2 in the gas phase requires a substantial activation energy, ~52 kcal/mole, and is a thermodynamically unfavourable process(173). In aqueous solution the activation energy is lower, 18 kcal/mole(174) but still high enough to constitute in principle a rate-limiting step in biochemical processes. However, in living cells the hydration and dehydration reaction:

$$H_2O + CO_2 \rightleftarrows H_2CO_3 (HCO_3^- + H^+) \qquad 5.19$$

is catalyzed by the enzyme carbonic anhydrase (CA), which is an extremely efficient catalyst with one of the highest turnover numbers known(175). The tertiary structure of the enzyme has been determined by X-ray diffraction methods(176) and several models for the reaction mechanism have been proposed(78). The enzyme and the reactions related to it have been studied with a number of kinetic and spectroscopic methods(78) and recently it has been the subject of several theoretical investigations(76, 77, 177, 178) limited to a very small portion of the active site.

The active site of CA appears as a relatively small cavity (~10 Å deep and less wide) of approximately conical shape, with the base of the cone towards the surface of the enzyme, and the vertex of the cone pointing to the interior of the enzyme. Nearly at the vertex there is a Zn^{2+} bound to 3 histidine residues. In the vicinity of Zn^{2+}, there is a glutamic acid residue, which can be either in the neutral or in the ionic form, and which has been proposed to be important for the catalytic activity of CA. The cavity is sufficiently wide to accomodate several water molecules. No unambiguous experimental information about the number and/or positions of water molecules in the active cleft is presently available, making the determination of the water structure in the active cleft an important task. We have, therefore, performed Monte Carlo simulations (76) of the water structure in the active cleft of CA using analytical pair potentials obtained from quantum mechanical calculations.(77)

A simulation of the solvation of the whole enzyme, requiring several hundreds of water molecules, is not presented in these notes. The system was reduced to a tractable size by 'cutting out' the active cleft and the amino acids bordering it. In the Monte Carlo simulation and in the iso-energy contour maps 27 amino acid residues were included(179). The coordinates used for the enzyme were those obtained from the X-ray study (176).

The first information that may be obtained is a more precise shape of the cleft of CA. This can be accomplished by calculating iso-energy contour maps for several cross-sections perpendicular to the one axis of the active cavity. The iso-energy contour maps were obtained by calculating a two-dimensional grid in which, at each x, y coordinate of the water oxygen atom, the orientation of the water molecule was optimized. The requirement for calculating iso-energy contours is a knowledge of the intermolecular potentials. The potentials needed are those between water and Zn^{2+} and between water and the amino acids(79). The Zn^{2+} has been placed at the origin of coordinate system, thus the 'cone' of the active cavity begins at a z-coordinate of ~0 and extends down to x=~ -10 A (the approximate axis of the cone has been taken as z-axis). Further down the active cavity opens up in an irregular way reaching the surface of the enzyme. The iso-energy contour maps have been calculated for 6 parallel planes perpendicular to the z-axis and following the variation of the contours, it is possible to form a quantitative image of the shape of the cavity, as experienced by one water molecule (see Figure 42). It is noted that this shape does not have to coincide with the one obtained, for example, by attributing van der Waal's radii to the atoms of the residues: the hydrophilic and/or hydrophobic character of each protein atom relative to the water molecule is included in the analytical atom-atom pair potentials and therefore the representation provided by the iso-energy contours is a quantitative one. In addition, it is noted that different assumptions on the deprotonation of the amino acids will bring about differences in the cavity shape.

The iso-energy contour maps give information about the interactions in the cavity felt by one water molecule. In order to obtain knowledge about the actual solvent structure (for example, coordination numbers, probability densities) it is necessary to perform some kind of solvent simulation. In this work it was done by applying the Monte Carlo technique to a sample of 20 water molecules at 300K confined to the active cavity by a cylindrical box ranging from z = 0 to z = -22.0 A and with a radius of 4.5 A. The amino acids of the enzyme were considered as a solute and were kept at fixed positions. In some additional calculations the dimensions of the cylinder were varied with only minor changes in the final probability distributions. The cylinder is longer than the cavity. This was done in order to provide a water reservoir to fill the cavity. The radius of the box was chosen in order to avoid a solvation of the outside of the cavity, since this is an artificial surface created by cutting out only a minor part of the enzyme. A total of 430,000 configurations were generated, 350,000 for the equilibration and 80,000 for the calculation of statistical averages. The average water-water interaction energy in the cleft was -10.5 kJ/mol which, compared to the same value for pure water, - 35.6 kJ/mol(90), indicates that the water structure in the active cleft is notably different from that in bulk water.

In Figure 43 is shown the radial distribution function Zn-O(r) up to 5 A away from the Zn^{2+}. Upon integrating the first three peaks, corresponding to an approximate first solvation sphere, it is found that they contain 1 water molecule each, while the fourth peak corresponds to 4 water molecules. Another representation of the water distribution in the active cleft is provided in Figure 44, where the probability distributions of oxygen and hydrogen atoms are shown. The probability distributions are presented for volumes obtained by segmenting the cylinder with planes parallel to the base and at an interval of 1 A. From the distribution functions shown in Figure 44, it is possible to obtain the orientation of the water molecules in the cleft. The x and y coordinates are given in the figure while the z-coordinate has an uncertainty of the sampled volume thickness (1 A).

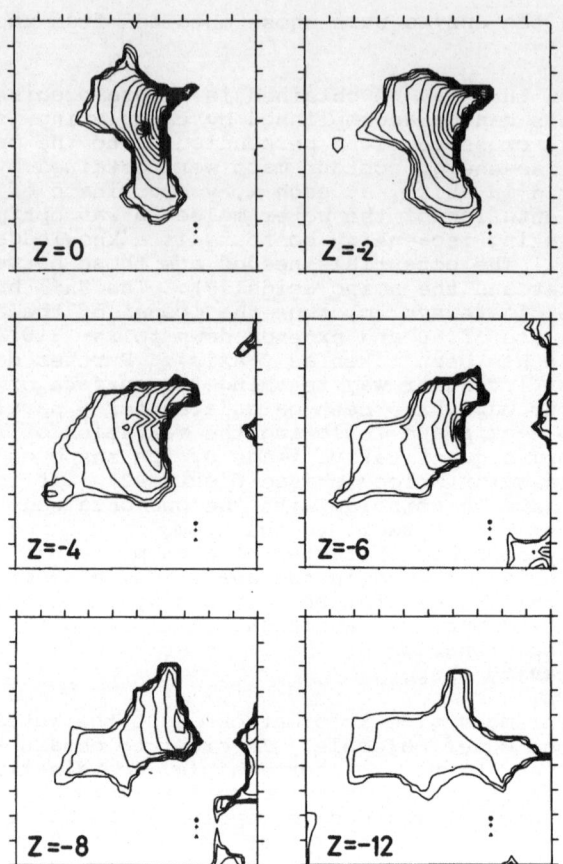

Figure 42. Iso-energy contour maps for 6 parallel cross-sections cutting through the cavity, parallel to the x, y-plane and at the indicated z-coordinates (a.u.). The contour-to-contour separation is 5 kcal/mole with the outermost contour corresponding to 0 kcal/mole. The separation between consecutive marks on the perimeter is 2 a.u. The Zn^{2+} defines the origin with respect to which the geometrical centre of each drawing has the coordinate $x = -3$ and $y = -1$ a.u. (1 a.u. = 0.529 Å).

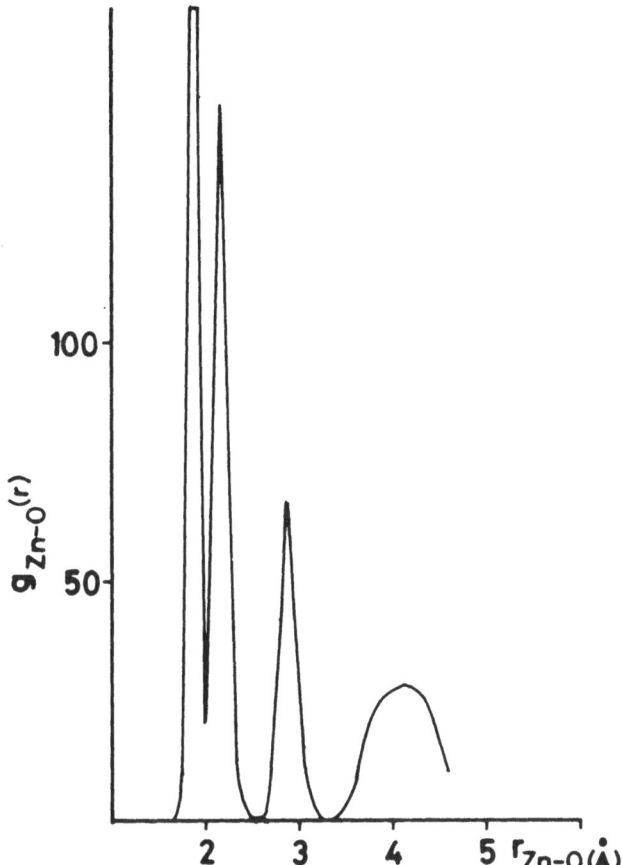

Figure 43. Radial distribution function, $g_{Zn-O}(r)$, for water oxygen atoms around Zn^{2+}, in the enzyme cavity. The first 3 peaks correspond to 3 water molecules and the following contains 4 water molecules.

The results suggest that the Zn^{2+} in CA is at least 5-coordinated and likely 6-coordinated(180). From the probability density plots it is also apparent that even several A away from the Zn^{2+}, there is a high degree of order in the water structure in the active cleft.

As known, the nature of the enzymatic mechanism is still uncertain, and a number of questions regarding the coordination number of the zinc ion, and as to how the CO_2 molecule is coordinated at the active site remain to be answered. In order to attempt to answer these questions, a series of Monte Carlo simulations of the water structure around Zn^{++} and Zn^{++}. CO_2 have been performed.

The structure of clusters consisting of 'n' water molecules around Zn^{++} has been obtained using MC techniques. The interactions have been limited to two body terms. The computed equilibrium distance was found

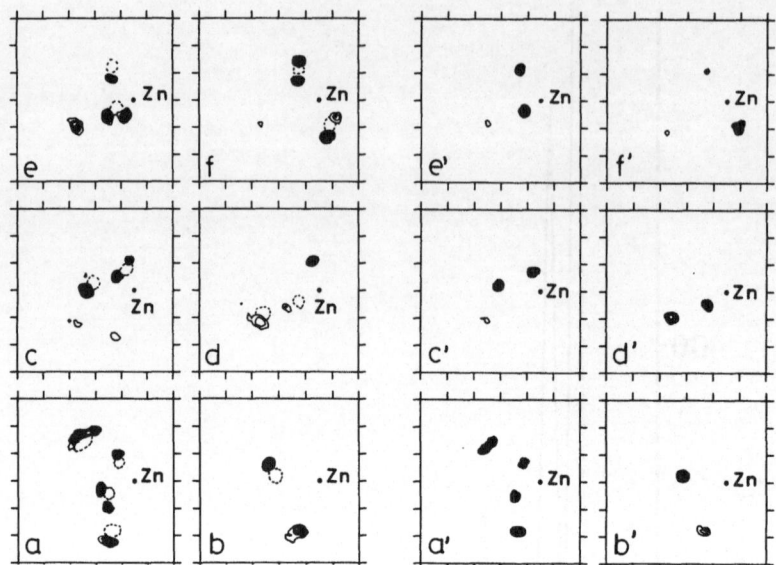

Figure 44. Probability density maps for water hydrogens (insert a-f) and oxygens (insert a' to f'). The maps correspond to cross sections through the cavity and parallel to the x, y-plane at the following z-values: a,a',z = -0.5 A; b,b',z = -1.5 A; c,c',z = 2.5 A; d,d',z = 3.5 A; e,e',z = 4.5 A; f,f',z = 5.5 A. In order to facilitate the reading of the maps, the oxygen maps have been added to the appropriate hydrogen maps (dashed lines).

to be 3.74 a.u. (zinc-oxygen distance) in a planar C_{2v}- symmetric conformation; the corresponding interaction energy was computed as -75.6 kcal/mole. Correction of the basis set superposition error with the counterpoise method changes the binding energy to -73.0 kcal/mole and the Zn-O separation to 3.85 a.u. The above computed binding energy and equilibrium position have been tested with an extended basis set (see Table 24) yielding the values -82.0 kcal/mole and 3.178 a.u. Previous computations (178) on the $Zn^{++}\cdot H_2O$ interaction have been reported with a binding energy of -112 and -104 kcal/mole; it is noted that this large binding energy is a consequence of the large superposition error.

In reporting the above binding energies, the dissociation products are assumed to be $Zn^{++}+H_2O$; as known, the correct dissociation products are $Zn^{+}+H_2O^{+}$. The $Zn^{++}\cdot H_2O$ species represents the inner complex, whereas the $Zn^{+}\cdot H_2O^{+}$ is the outer complex(155), the latter being more stable than the former.

The computed interaction energies have been used to obtain analytical pair potentials of the standard form (see eq. 4.2).

The values of the fitting constants, A, B, and C, and the net charges are given in Table 42. The standard mean square deviation, 3.3 kcal/mole, is less than the estimated error in the computed binding energy.

The interaction potential between CO_2 and H_2O has been obtained in an

Table 42. Fitting parameters in the intermolecular potentials for $Zn^{++} \cdot H_2O$ and $CO_2 \cdot H_2O$. The net charges on water oxygen and hydrogen were -0.6836 and 0.3418, respectively. The total energies for the monomers were: Zn^{++}=-1769.1770, H_2O = -75.7427 and CO_2 = -186.9337 a.u.

	With H			With O			
M.	A	B	C	A	B	C	q
Zn^{++}	263.886	0.112959	1.37483	-988.184	5537.40	1.39417	2.0
C(CO_2)	0.000925	11873.1	1.00698	549.867	53434.3	1.00728	0.794
O(CO_2)	10.7055	3128.68	0.995753	0.000853	155516.	0.997514	-0.397

Mean square deviation (kcal/mole) for $Zn^{++} \cdot H_2O$ is 3.3, for $CO_2 \cdot H_2O$ is 0.4.

analogous manner. Details of the interaction potential are found in Table 42. The total energies, coordinates and interaction energies for the two intermolecular potentials are available elsewhere(179).

A number of water clusters of different size (n = 4, 5, 6, 7, 8, 9, 10, 50) have been analyzed (the number of moves in the different MC-simulations were n4000 for equilibration and n4000 for the calculation of the statistical averages), and the results are summarized in Table 43. Recalling that the $Zn^{++} \cdot H_2O$ attraction is ten times as large as the water-water interaction in bulk water, one would expect a number of solvation shells around the zinc ion; the orientation of water molecules especially in the innermost shells is expected to be determined by the $Zn^{++} \cdot H_2O$ interaction at the expense of the water-water interaction, as is indeed found (see Table 43). The zinc-oxygen radial distribution function, $g_{Zn-O}(r)$, has been computed. In the case of fifty water molecules, two distinct peaks appear in the radial distribution function (see Figure 45), which can be used to define shell radii. In the two shells the dipole moments are essentially parallel to the line joining the oxygen atom with the zinc ion. The definition of shell radii is not unique, since the hydrogen atoms of a water molecule are farther from the zinc ion than the oxygen atom of the same water molecule(156, 157). To define a third solvation shell, a large sample than fifty water molecules would be needed. The zinc-water interaction in the first shell is ~-69 kcal/mole, that is, a little less than the binding energy of the $Zn^{++} \cdot H_2O$ complex at equilibrium. In the second shell, the binding energy drops sharply, but is still about twice as large as the water-water interaction in bulk water (90).

Table 43. Water clusters containing Zn^{++}. All energies in kcal/mole.

Number of H_2O molecule	4	5	6	7	8	9	10	50
Water-Water interaction(a)	5.00	7.55	9.25	8.34	7.43	6.78	6.35	-1.22
Water-Solute interaction(b)	-280	-346	-410	-436	-463	-486	-507	-816
Total energy(a)	-65.1	-61.6	-59.1	-54.0	-50.3	-47.3	-44.4	-17.5
Number of water molecules in the first two shells	4,0	5,0	6,0	6,1	6,2	6,3	6,4	6,16
Water-Water interaction in the first shells	5.00	7.55	9.25	9.72 0.15	9.94 0.15	10.18 0.15	10.30 0.6	11.3 -0.6
Water-Solute Interaction in the first shells	-70.0	-69.1	-68.3	-68.3 -26.7	-68.2 -26.0	-68.3 -25.3	-68.3 -24.2	-68.7 -17.4

(a) The statistical error is ±0.1 kcal/mole.
(b) The statistical error is ±4 kcal/mole.

In the active cleft of carbonic anhydrase, carbon dioxide must diffuse towards the zinc ion. Since the active cleft is capable of accommodating several water molecules, it is of interest to analyze the solvation of a system consisting of Zn^{2+}, CO_2 and several water molecules.

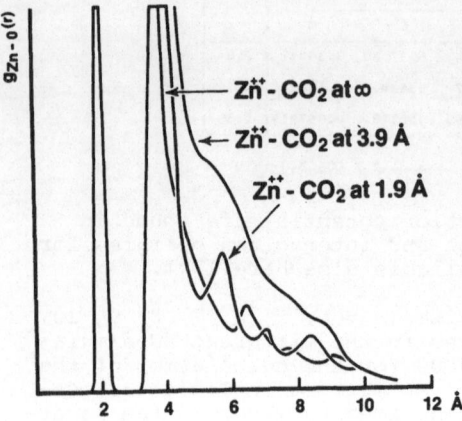

Table 44. Water clusters containing Zn^{++} and CO_2 at 1.9 Zn–O separation. All energies in kcal/mole.

No. of H_2O Molecules	10	50
Water-water energy	4.6±0.1	-1.4±0.1
Water-Solute Energy	-458±0.5	-778±2
Total Energy	-41.3±0.1	-16.9±0.1
Number of Water Molecules in the first two shells	5,5	5,13
Water-Water interaction in the first two shells	+8.3 0.9	+9.4 +0.15
Water-Solute Interaction in the first two shells	-66.4 -25.2	-66.7 -20.1

Figure 45. Radial distribution functions for water oxygen atoms around the zinc ion. Number of water molecules = 50.

Preliminary simulations have been performed with fifty water molecules considering the $Zn^{2+} \cdot CO_2$ complex as a rigid solute, only the water molecules being free to adjust their position. The MC simulations were performed at T = 300 K and the volume of the Monte Carlo box was $32^3 \text{ Å}^3 = 32768 \text{ Å}^3$. The system was equilibrated with 200,000 configurations. The main results are summarized in Table 44 and in Figure 45. The distribution of water molecules in the third solvation shell has not been analyzed, as it has little statistical value. For the second shell, it is noted that the second peak of $g_{Zn-O}(r)$ is somewhat shifted away from the zinc ion in the case of the $Zn^{++} \cdot CO_2$ complex. From the data in Table 44, it is seen that when CO_2 is placed at 1.9 Å (its minimum energy position in the $Zn^{++} \cdot CO_2$ complex) it displaces one water molecule, yielding a first solvation shell with five water molecules and one CO_2 molecule. The water-water interaction in this case is only slightly more attractive than in the solvation of Zn^{++} only, in agreement with the observed shift of the peak in $g_{Zn-O}(r)$.

The coordination number of Zn^{++} is commonly determined as 4 or 6, depending on the ligands. Recently the water coordination number has been determined with X-ray methods(181), and found to be 6, in agreement with the theoretical result. The first peak in the experimentally determined radial distribution function was found at 2.08 Å in close agreement with our theoretical value of 2.0 Å.

In the radial distribution function (Figure 45 curve b) it is seen that the first peak is very sharp and well defined, indicating that the ion and the first solvation shell form a rather rigid structure.

It is interesting to note that introducing a CO_2 molecule in the first solvation shell displaces one water molecule, apparently without significantly altering the remaining water structure in the first shell. The effect of the presence of the CO_2 molecule on the second shell is less significant statistically, but it seems as if the CO_2 molecule also displaces some water molecules from this second shell, causing the remaining water molecules to interact more strongly with the zinc ion. These results are not unreasonable when the shape of the CO_2 molecule is considered.

In order to obtain macroscopic quantities such as a solvation energy of Zn^{++} it would be necessary to apply some type of periodic boundary conditions in the MC simulations, and this has not been done in this work. However, it is possible to obtain an estimated value for the solvation energy by considering the remaining space, outside the water cluster, as a continuum. Applying the Born model(182), the solvation energy for the $Zn(H_2O)^{++}$ cluster is found to be -542 kcal/mole, whereof the continuum contribution is -92 kcal/mole. In obtaining this value the volume of a water molecule and of a zinc ion have been set equal to 30 A^3, giving the radius for the cluster. The dielectric constant for water has been set to eighty. The agreement with experiment, -494 kcal/mole for the enthalpy of solvation(183), is fairly good and shows that the calculated $Zn^{++} \cdot H_2O$ potential is of reasonable accuracy.

As previously noted, the use of only two-body interactions underlines the preliminary nature of this simulation. These studies, being preliminary, are being continued; it is a pleasure to thank Prof. I. Bertini (Florence, Italy) and Dr. S. Koening (IBM Research, Yorktown, New York, U.S.A.) for stimulating discussions.

5.10 CONTOUR MAPS FOR THE MOLECULAR FRAGMENTS OF DNA

The analytical pair potentials we shall use have been constructed to reproduce, with a mean standard deviation of about 5%, a large number of intermolecular interaction energies obtained from ab initio computations.

In Figure 46 (right inserts) we report the isoenergy contour maps for a model compound containing a $-PO_4^- -CH_2-$ group terminated at one end with a $HC-(CH_3)_2$ group and at the other end with a $-HC(CH_3)(OH)$ group(184). In this model compound the nearest neighbor atoms of the $-PO_4^- -CH_2$ unit are located as in DNA. We have considered six planes parallel to the plane containing the atoms O1, O2, and P: our x, y reference plane is at z=0. The six planes are at $z=\pm 1$, $z=\pm 3$, and $z=\pm 6$ a.u.; the grid interval selected to obtain isoenergy contour maps is of 0.5 a.u. The area considered for each plane is rectangular with length 36.7 a.u. along x, and 33.3 a.u. along y. The distance between two successive marks on the perimeter of the inserts in Figure 46 is 3.3 a.u. As it was previously found for diethylphosphate(185), in the isoenergy contour maps for the model compound, we can distinguish four regions: a strong hydrophilic region (region a, see top right insert for Z=1 of Figure 46), a second hydrophilic region, weaker than the previous one (region b), a hard-core (region c), and an hydrophobic region (region d). By increasing the distance z from the reference plane, regions a and b merge into a single region centered on the two terminal oxygen atoms O_1 and O_2. We note that in DNA and RNA, the two terminal oxygen atoms, are at the outmost periphery of the helix. Therefore, for each phosphate group in DNA (and RNA), we expect to find the equivalent of region a; region b is expected to be rather different because of the interference of the sugar-base units.

In Figure 46 (left insert) we report the isoenergy maps for the sugar unit(75). The reported maps are selected for planes parallel to the molecular plane containing the atoms C1, C2, and C3 and representing x,y plane with z=0. The planes we considered here for the maps are at z=0, ± 2, and ± 4.0 a.u. Note, the variations in the hardcore with z, the hydrophilic regions due to the ring oxygen and to the OH group. By comparing the area included in the hydrophilic region in the phosphate, in

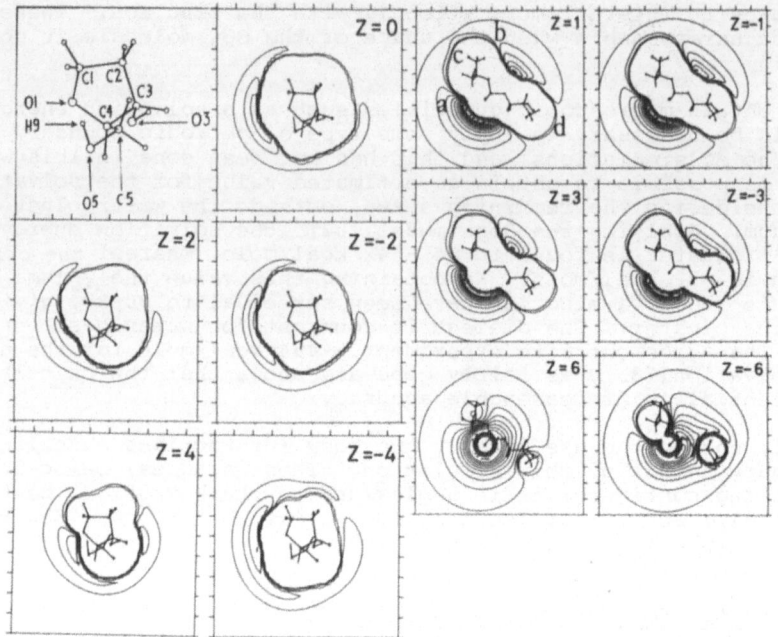

Figure 46. Iso-energy contours maps for sugar and model compound containing the PO_4 - group. Map to map displacement (z) in A.

the bases, and in the ribose derivative, one can expect that the dominant feature of DNA in solution is related to the hydrophilic region of the phosphate groups.

In Figure 47 (top inserts) we report the five bases of nucleic acids indicated with a, c, g, t and u; the molecular geometries are those given in (186). The five inserts of Figure 47 report isoenergy contour maps obtained (for the plane containing the base) by considering the interaction energy of one water molecule with the base at T = 0 K. For a few selected positions we have explicitly indicated the orientation of the water molecule, that by construction is constrained to have the oxygen atom in the molecular plane. The contour to contour interval is 1 kcal/mole; the innermost contour is 1.0 kcal/mole (nearly super-imposed on the contour at 0.0 kcal/mole, not given); the outermost contour is -1.0 kcal/mole.

In Figure 47 (bottom inserts) we report the isoenergy contour maps for molecular planes containing the base-pairs A-T, G-C and A-U; these are generated with a resolution of 0.20 A and cover an area of 20 x 16 A^2. The hydrogen atoms forming base-base hydrogen bonds, have been kept at the same distances as used in the single base study. For a discussion of the base-base hydrogen bond energy we refer to an earlier paper (187) and to a recent review (188). In Figure 47, the contour to contour intervals correspond to an energy difference of 1.0 kcal/mole. The innermost contour is the -1.0 kcal/mole contour (superimposed on the 0.0 kcal/mole contour, not given). A few water molecules are explicitly given and correspond to the water molecules at a position near the molecular plane, determined with MC method later in this chapter. The water mole-

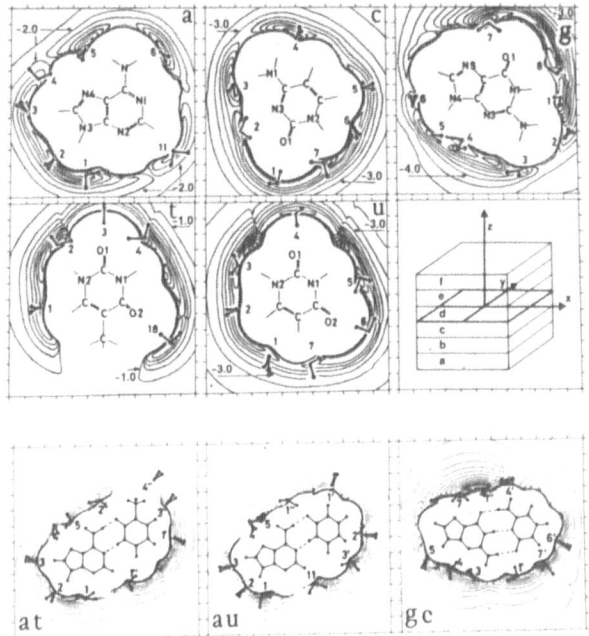

Figure 47. Iso-energy contour maps for bases and base-pairs.

cules are designed with the same numerical code used later in this work. Unprimed numbers are used for water molecules that are found at nearly the same position and orientation as determined for one of the two bases of the base-pair; primed numbers are used for water molecules equivalently related to the second base; double prime numbers are used for water molecules that hydrate the base-pair but cannot be identified with any of the hydration sites in the separated bases (see later sections).

As one can notice, some of the water molecules are at energy minima, some are not: this is because in the isoenergy maps the solvent-solvent interaction and the temperature effects are ignored. Whereas, the position of oxygen atoms in Figure 47 is the one determined by MC, the orientation of the water molecules are those obtained in the orientation optimization step during generation of the isoenergy maps. The isoenergy maps are accurate to about ±2.0 kcal/mole, relative to S.C.F. ab-initio computations; in turn, these computations are expected to be over estimated by 1 to 2 kcal/mole relative to more accurate quantum-mechanical computations.

The Figures 46 and 47 refer to the units of nucleic acids. In Figure 48 we consider the isoenergy for a fragment of A-DNA single helix (189, 190) reported in Figure 48 (left insert).

The atomic coordinates for this conformation are those deposited in the N.I.H. Atlas of Macromolecules (191); the coordinates for the hydrogen atoms have been added with the aid of a special computer routine making use of previously reported bond lengths and bond angles for the hydrogen atoms in the four bases (186,187) and in the sugar unit (41).

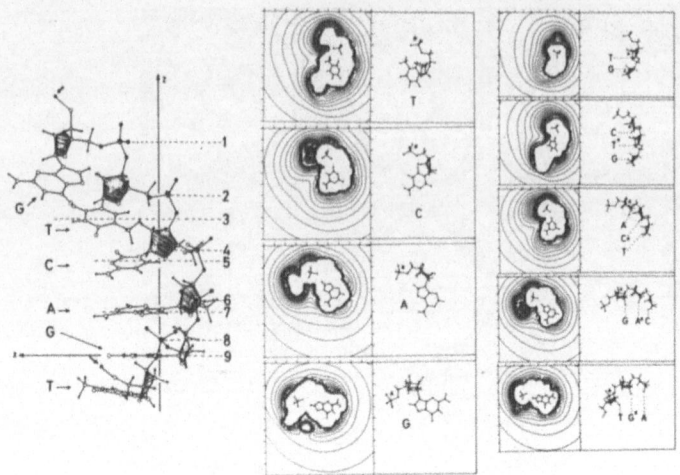

Figure 48. Three-dimensional projection of A-DNA single helix fragment (left) iso-energy maps for planes 2, 4, 6 and 8 (center) and remaining planes (right).

We have considered six of the repeating units (sugar and base -- PO_4^- -- CH_2 --) with the following sequence for the bases; G, T, C, A, G, and T. Our main interest is in the central four units -- T, C, A, and G; the first and last unit (with G and T, respectively) are included only to ensure proper boundary conditions in the computation of the interaction between water and the above DNA fragment composed of 191 atoms (Figure 48).

The first sugar unit is terminated with an H atom placed at a position corresponding to a P atom for a more extended fragment. The last sugar unit is followed by a -- CH_2 -- group terminating with an OH group whose oxygen and hydrogen positions correspond to an oxygen and the phosphorus

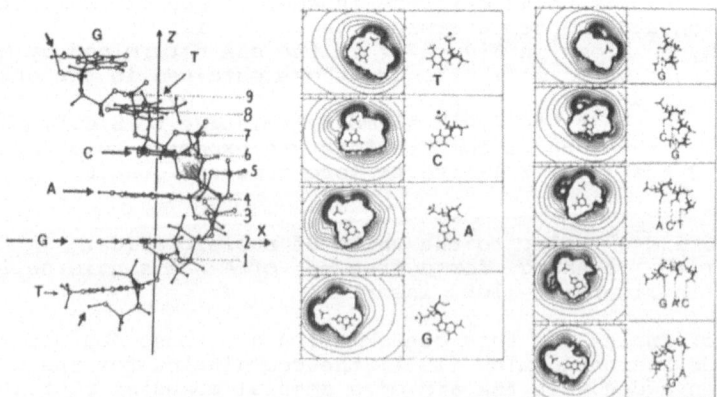

Figure 49. Iso-energy maps for B-DNA single helix (see Figure 48).

position of a PO_4^- group for a more extended fragment. The terminal hydrogens at the two ends of the fragment are identified in Figure 48 with an arrow. We note that the exact position of the two terminal atoms is of no consequence for the results reported here.

For B-DNA single helix (192,193) we consider an equivalent fragment. The relative position of the bases in the B-DNA single helix fragment is that given in Figure 49 (left insert).

We present four maps (Figure 49, middle inserts) obtained by computing the interaction energy of a single water molecule with the 191 atoms of our fragment of B-DNA. The maps are selected in the x, y plane at variable z; the origin of the Cartesian system is at the nitrogen atom bonded to the sugar in G (see plane 9 in the left insert of Figure 49); the guanine molecular plane defines the x, y plane in our system (the coordinates presented in Reference 193 have been rototranslated; the original coordinates are used later in this paper). The maps reported in Figure 49 (middle inserts) are labeled T, C, A and G and correspond to the planes 8, 6, 4 and 2 of the left insert of Figure 49. On the right hand side of each map we have reported a three-dimensional representation not only of the phosphate and the base in the proximity of the map's plane, but also of other groups near the phosphate or the base. The area covered by each map is 52 x 52 a.u., namely, each map is constructed by computing the interaction energy with DNA (at the optimal orientation) for 2704 molecules of water, considered one at a time. The contour-to-contour energy interval is 2.0 kcal/mole; the marks on the perimeter of the maps correspond to a segment of 5.1 a.u. length. In the four maps we can readily indentify the hard-core region containing a base and part of a sugar and a phosphate. When the base is not exactly in the x, y plane the hard-core region is somewhat more extended since more of neighboring atoms are included. The inclusion in our fragment of DNA of the two terminal repeating units (sugar-G phosphate) and (sugar-T phosphate) is clearly needed in order to obtain a realistic representation of T and G (planes 8 and 2, respectively, in Figure 49). Note that the hydophilic region is very strong and extended with many well defined minima. As expected the deeper minima are in\ the vicinity of the phosphate group where we can count up to 15 contours (corresponding to 30 kcal/mole).

In Figure 49 (right inserts) we report five maps for planes parallel to those of Figure 49 (middle inserts), but passing through the phosphorous atom of the five phosphate groups labeled 9, 7, 5, 3, and 1 in Figure 49 (left insert). The dimension and contour-to-contour interval of these maps are equal to those discussed above. The hard core region for map 9 is smaller than the corresponding hard core regions of the other maps reflecting the proximity of the fragment's end; for planes 7, 5, 3, and 1 one can see the hard core corresponding to the bases T, C, A, and G, respectively. The hydrophilic region extends over the entire map (of course, the hard core is excluded); the attraction is very strong near the phosphate groups. By considering the maps of Figure 49 one can see an overall rotation of the hard core following the helix main axis; thus we can obtain a representation of the "shape" of DNA as seen by water.

In general, contour maps are reported for selected planar cross-section. From the point of view of a study of DNA in solution, however, it is probably more interesting to obtain maps corresponding to cylindrical surfaces enclosing the DNA helix. In Figure 50 (top inserts) we present schematically such cylindrical areas. The radius of the circumference of the planar cross-section perpendicular to the long helix axis (z axis) and containing the phosphorus atoms in 16.86 a.u. for A-DNA and 16.84 a.u. for B-DNA. A number of concentric cylinders have

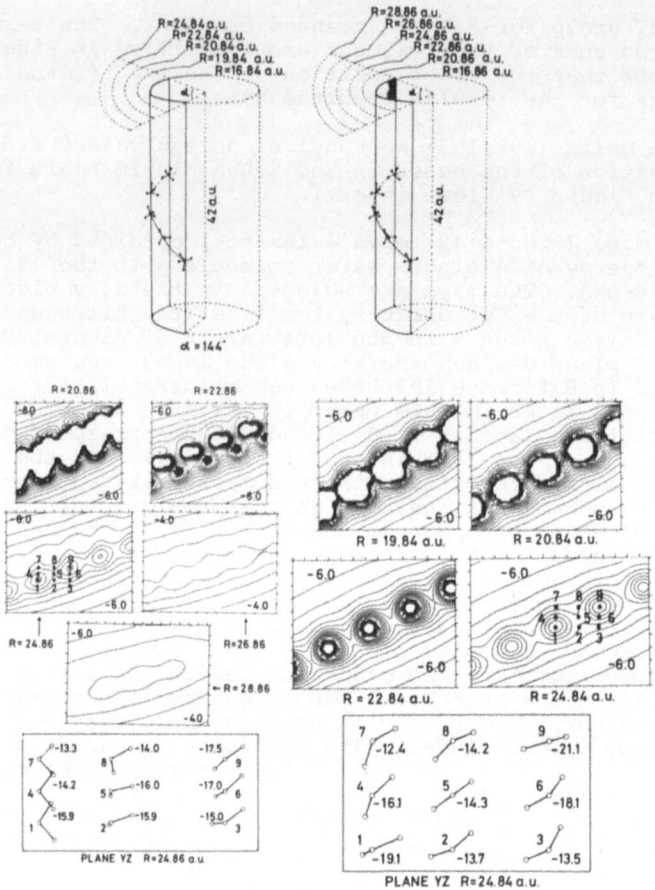

Figure 50. Cylindrical iso-energy maps for A-DNA (left) and B-DNA (right). The radii selection is indicated in the top inserts. The optimal orientation and energy (in kcal/mole) for a sample of water molecules is given in the bottom inserts.

been considered. For each cylinder we have selected an area with 42 a.u. along the z direction and of width corresponding to an arc α of 144° which includes five consecutive phosphorus atoms, as schematically reported in Figure 50 (top inserts). The maps corresponding to such cylindrical areas are presented in Figure 50 (left inserts for A-DNA and right insets for B-DNA). The results are both suggestive and self explanatory. Progressing from the innermost plane to the outermost plane the hard core regions disappear and the hydrophilic contours become gradually less dense. We notice, as expected, (a) a very high periodicity, (b) strong hydrophilic minima centered at the phosphate groups, and (c) a very extended area of intermediate hydrophilic region (see next section). At the top and bottom of each map we provide the energy value of the outermost and less energetic contour. By comparing A-DNA (map at R - 20.86 a.u.) and B-DNA (map at R = 20.84 a.u.) one notices the change in the hard core size and in the isoenergy contour density, indicating notable differences in the hydration of the two forms of DNA.

Notice also the importance of cooperative effects. Indeed, remembering that in our computation there are two terminal (sugar-base-PO_4^-) units, one can notice (see map at R = 28.86 a.u. for A-DNA) that the innermost contour encloses three repeating units but not the first and the last one. This constitutes a strong hint that the total field at one unit would be incorrectly computed, if one neglected the field of the two meighboring units.

The same for the other maps, but can be seen less readily because of the closeness of the contours. Let us now consider in detail the position and energy of a few molecules of water for two maps. We select nine positions for molecules of water located in the vicinity of the phosphate minima. At the bottom of Figure 50, we report a three dimensional representation of these nine molecules of water and the corresponding binding energies. The directionality and periodicity provides a most interesting pattern, that is repeated from unit to unit. The differences in the two DNA forms, both in orientation and energy, again indicate rather sharp differences in the solvation of the two species.

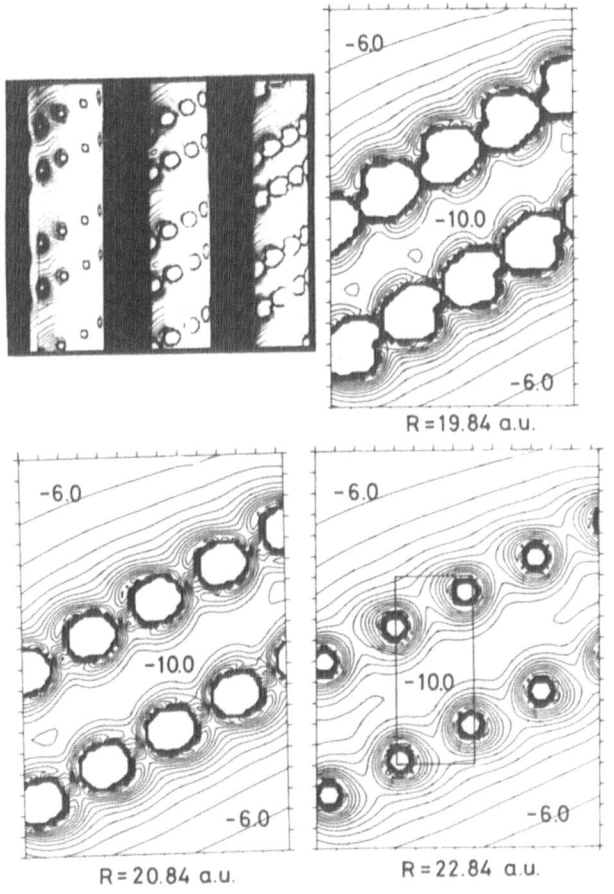

Figure 51. Cylindrical iso-energy maps for B-DNA double helix.

In Figure 51 we report three cylindrical maps obtained by computing the interaction of water with the double helix of B-DNA for R = 19.84, 20.84, and 22.84 a.u. (For greater details, we provide elsewhere (192) six cylindrical maps corresponding to the area schematically indicated by the rectangular area in the map at R = 22.84 a.u.). The results of these maps are partially expected on the basis of Figure 50; however, the main unknown in passing from a single B-DNA helix to the double B-DNA helix is the strength of the attraction exerted on water in the minor groove region.

It is important to notice that in this region the interaction is approximately constant (between -12 and -9 kcal/mole); a substantially higher value, as in the single helix, would have been in disagreement with the problem of explaining the relatively easy transitions in solution to a single helix from the double helix.

Previously, we have analyzed the water structure around A-DNA and B-DNA, confining our attention in the vicinity of the phosphate groups and outside DNA (189,190,192). In the following we extend the analyses to water molecules located in the major and in the minor groove of B-DNA double helix. The computations (194) are of preliminary nature since they are not Monte Carlo simulations but only the isoenergy maps.

DNA can be enclosed into a cylinder with its axis (Z axis) coinciding with the long axis of DNA. In B-DNA double helix the phosphorous atoms fell on a cylindrical surface of 8.91 A radius. For DNA in water solution, due to the strong attraction of the phosphate groups with water molecules and to the location of the phosphate groups at the perifery of DNA, we can consider a cylinder of radius R = 10 A as delimiting two regions of the solvent, one with R > 9, for bulk-water and the hydration of the PO_4^- groups, the second with R < 9 for water molecules penetrating DNA and experiencing the filed of the sugars and the base-pairs.

In Figure 52 we note the number of water molecules inside the minor and major groove (see maps at R = 4, 5 and 6A) is smaller than the equivalent number of water molecules at the periphery of DNA (see map at R = 9 and 10A), and as a consequence the solvent-solvent interaction is less dominant. From the isoenergy maps analysis, we can confirm the intuitive ideas related to hydration in the major and minor groove; we, however, add new information, namely we quantitatively determine how near to a given position of the base-pair a water molecule can go, balancing the strong field of the phosphate groups and the field of the base pairs and sugar fragments. Finally, from the iso-energy maps reasonable guesses can be extracted concerning the "effective space" available to molecules complexing or interacting with DNA.

The structural data used in this work are those reported elsewhere(194). We have selected a fragment of B-DNA double helix, that includes twelve base-pairs and the corresponding sugar-$CH_2PO_4^-$ groups for a total of 760 atoms; the fragment extends for 40 A (Z direction). The base-pair sequence is given in Table 45; for each base-pair we have indicated an average <Z> value (the base-pairs have the molecular plane not exactly perpendicular to the Z axis). The first base of a base-pair in Table 45 belongs to one helix, the second base to the second helix in the double helix. In Figure 53 we report three dimensional projections for the base-pairs (and neighboring groups) in Table 45, excluding two pairs, the first and the last one. In the following of this work, isoenergy contour maps are not reported for regions corresponding to the first and last pair of Table 45, to eliminate discontinuity effects due to the head and tail of the DNA fragment; however these base-pairs and related terminal groups are included in the computations of the interaction energy here reported.

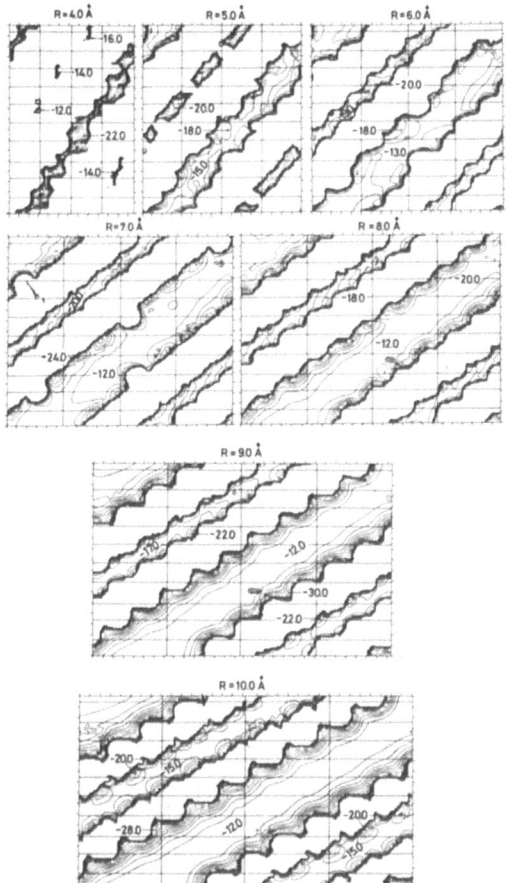

Figure 52. Cylindrical iso-energy maps for selected areas of B-DNA double helix.

In Figure 52 we report the isoenergy contour maps obtained for cylindrical surfaces of R=4.0, 5.0, 6.0, 7.0, 8.0, 9.0 and 10.0 A. The ordinate (Z direction for the DNA fragment) extends from 4.0 A to 38.4 A; the abscissa extends from 0 A to $2\pi R$ A. The horizontal lines give the <Z> values for base-pairs as in Table 45; the vertical lines subdivide the circumference into quadrants starting from the -X axis; the selected rotation is counterclockwise. The interval between two successive marks on the borders of the maps corresponds to 1.9 A. The contour to contour energy difference is 2.0 Kcal/mole.

We have added circumferences to Figure 53 corresponding to the isoenergy maps at R=4.0, 6.0, 8.0 and 10.0 A, given in Figure 52; with the combined use of Figure 52 and 53, one can identify the interaction energy for a water molecule with DNA for the entire region 4.0 A <R<10.0 A; for R>10 A we refer to our previous works(189, 190 and 192).

The use of Figures 52 and 53 is exemplified for the case of one base pair.

Table 45. Sequence of base-pairs in the fragment of B-DNA and relative positions (in A).

#	Pair	<Z>	#	Pair	<Z>	#	Pair	<Z>
0	A-T	40.4	4	C-G	27.1	8	A-T	13.4
1	T-A	37.2	5	T-A	23.8	9	T-A	10.2
2	G-C	33.7	6	G-C	20.1	10	C-G	6.8
3	A-T	30.3	7	C-G	17.0	11	G-C	3.2

Let us consider, as an example, the fifth pair, T-A of Table 45: the XY plane containing this base-pair, corresponds to the fifth horizontal line (counting from the top) at Z=23.8 A (see Figure 53). At R=4.0 A, near the end of the first quadrant (that is, near the -Y axis of Figure 53 or near the first vertical line of the insert R=4 A, in Figure 52) and the third quadrant (that is, near the +Y axis of Figure 53 or near the third vertical line of the insert at R=4 A, in Figure 52) we find attractive regions of about -12 Kcal/mole and -14 Kcal/mole, respectively. These

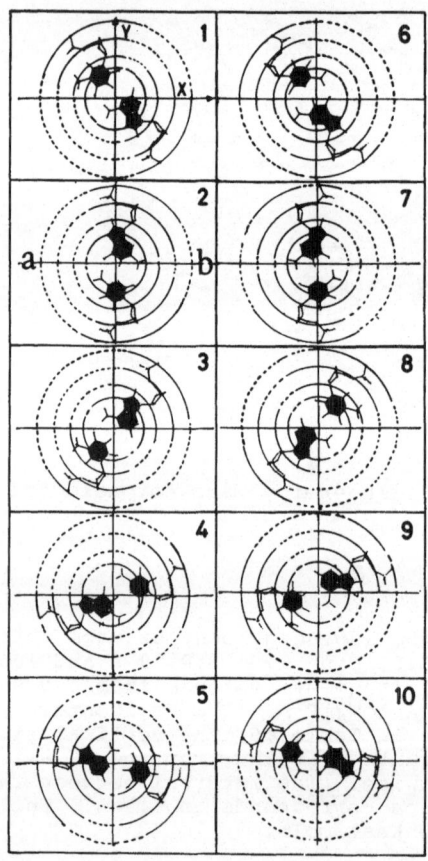

Figure 53. Ten base-pairs of B-DNA double helix fragment (two additional base-pairs are not reported in the figure, see text).

regions correspond to hydration sites for the CO group in thymine, (shifted by the perturbation of adenine) and for the NH_2 group in adenine. In the Monte Carlo simulations of the base-pairs (see later), we have identified these hydration sites with two water molecules, W(1") and W(5). From the isoenergy maps there is no evidences of a site corresponding to W(2"), possibly because here we ignore both temperature and solvent effects. At R=5 A (see second insert of Figure 52) the intersections of the fifth horizontal line (that is, $<Z>$=23.8 A) with the isoenergy contours, indicate the existence of two regions with rather deep minima; these regions are present also in the map at R=6 A and are deeper than those reported for the base-pair or reported earlier for the individual bases. This increase in the attraction at large distances from the base-pair is due to the field of the phosphate groups (as can be seen by considering the remaining maps of Figure 52).

This type of analysis can be carried on for all ten base-pairs of Table 45; the qualitative aspect is summarized in Figure 53, where we indicate with dotted lines regions of attraction between B-DNA and a water molecule and with full lines we indicate the repulsive (hard core) region; the circumferences in Figure 53 provide, therefore, a condensed but only qualitative view of the hydration inside B-DNA; the corresponding energetic quantitative details are given in Figures 52.

The dotted and full line portions of each circumference in Figure 53 indicates that for each base-pair we can identify three regions. One region is characterized by repulsive interaction towards a water molecule (hard core region). Two regions (separated one from the other by the hard core region) are both attractive to a water molecule, but as expected, one is more extended than the other. These two regions are indicated by the letters "a" and "b", repsectively, in the second insert of Figure 53. The hard core region is due to the base-pair and corresponding sugar-CH_2-PO_4^- groups and to the two base-pairs immediately above and below the one considered. The different extension of the "a" and "b" regions is due to the fact that in the base-pairs, one side of the pair is near to the second helix of the double helix (minor groove), the second side is relatively far and, therefore, can accomodate more water molecules (major groove).

A second characteristic is that the hydration sites for the base-pairs in the region from R=4 A to R=6 A and coinciding with the first hydration shell for the separated base-pair in solution have approximately the same energy as in the separated bases. However, the second solvation shell (from R=6 A to R=10 A) has a solvation energy _larger_ than the one in the first shell because the attraction of the phosphate groups. More quantitatively, from Figure 52 we learn that the attraction water B=DNA double helix at R=4 A and R=6 A is of 10 to 14 kcal/mole, then increases to about 15-18 kcal/mole at R=8.0 A and to 20 Kcal/mole at R=10 A; after reaching a maximum of attraction, decreases (189 and 192). Thus, there is a very strongly bound water network between R=10 A and R=12 A that become weaker both at smaller and larger R values. A variation in the hydration network in this region affects the B-DNA structure, and conversively, different hydration structure correspond to different conformation of DNA. As known, variations in the solvation structure can be induced by the presence of different cations which bind to the phosphate and rearrange the hydration shells structure; another mechanism is related to temperature variations. As pointed out, at T=300 K the probability distribution for water molecules, obtained by Monte Carlo simulations, extends over a rather large volume for a given water molecule; in the first solvation shell the oxygen atom (of a water molecule) can move in an interval extending up to 2 A and at very little energy expense.

This motion can be concerted with equivalent motions of other water molecules in a strong binding region (189 and 192). From analysis of the isoenergy maps, we tentatively postulate that water molecules in the region from R=9 A to R=12 A could undergo concerted motions exerting a torque on the DNA double helix; the asymmetry of the "a" and "b" hydration regions (see Figure 53) tends to enhance the torque effect.

Despite the preliminary nature of the analysis reported(194) for B-DNA, we feel that the data implied in Figure 52 did represent the first reasonable attempt to theoretically predict the forces acting on a water molecule from the full field of DNA in a double helix conformation in the region near the base-pair.

5.11 MONTE CARLO SIMULATIONS FOR BASES AND BASE-PAIRS IN NUCLEIC ACIDS

A cluster of either forty or fifty molecules has been used to simulate the structure of the first hydration shell for the bases and base-pairs of nucleic acids (194). The MC results are analyzed with the aid of probability distribution (PD) maps; in these maps the interval between two successive contours is 0.2 atoms/A^3.

The molecular plane of the solute is assumed as the X,Y plane with Z=0. The probability distribution is analyzed in a volume with a MxN A rectangular base and 6 A height. To obtain a finer representation, this volume is subdivided into six subvolumes designated with the letters a to f, each with 1 A height, as shown in the bottom-right insert of Figure 47. The base is 12x12 A for the nucleic bases, and 16x24 A for the base-pairs.

The hydrogen probability distributions (PDs) are presented as separate from the oxygen PDs. The diffuse nature of PD is due to two factors, namely the near degeneracy for many water conformations, present even at T=0 K, as previously pointed out (70) and the thermal broadening (161, 163, 189, 192). The specific shape of the volumes enclosed by PDs is of particular interest; whereas a nearly spherical shape is an indication that the water molecule is in a nearly isotropic field, elongated shape indicate a preferential direction of the interaction field intensity due either to the solute or the rest of the solvent.

The isoenergy contour maps provide a clear picture of such directionality (see for example in Figure 47 the depth and length of the valleys on the side of -NH$_2$ groups). For the case of the bases in solution, the attraction due to the base on a water molecule is at most 5 to 7 Kcal/mole larger than the attraction of the remaining of the solvent on the same water molecule. Therefore, it is clear that the water-water interaction has a prominent role in determining the water structure in the first hydration shell; as a consequence, we cannot expect that the most probable hydration sites will always coincide with the minima of the isoenergy contour maps. Indeed, this can be noticed by considering the water molecules explicitly represented in Figure 47, that correspond to the most probable positions obtained in the MC simulation as discussed below. One can see that not all water molecules are at energy minima. The orientation of the water molecules corresponds to the best orientation for a single water molecule. Later we shall see that only this orientation seldom remains the most probable one when in solution at T=300 K. For a detailed discussion on the relationship between isoenergy contour maps and MC simulations we refer to Section 5.5.

In Figure 54 we report the PD maps for the oxygen atoms (inserts a to f on the left) and for the hydrogen atoms (inserts a to f, on the right) for the water molecule around adenine at T=300 K. In the oxygen maps the most probable water molecules positions are labelled with a number from 1 to 17. The marks on the perimeter of each insert refer to a segment 2 A in length. When the same water molecule has a finite PD of being in more than one insert, the numerical designation is given in one of the inserts only.

Five water molecules hydrate the NH$_2$ group, two in the plane, W(5) and W(6), one above the plane, W(13) and two below, W(10) and W(16). Above the plane, W(9) partly hydrates the NH$_2$ group, but the main interaction is with the N(4) atom.

Four water molecules hydrate the N(3)-H group, two in the plane, W(1), strongly perturbed by N(2) and W(2), one above the plane, W(8), and one below the plane, W(7). The lone pairs of the three nitrogen atoms

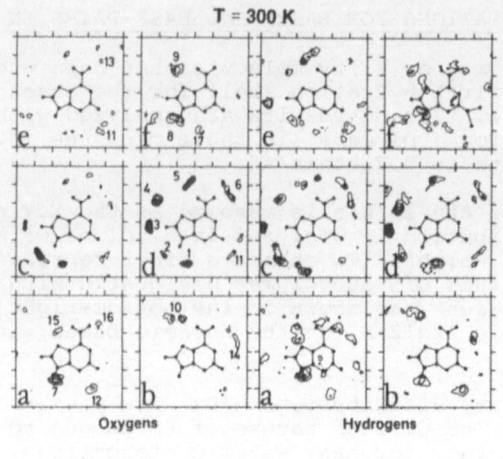

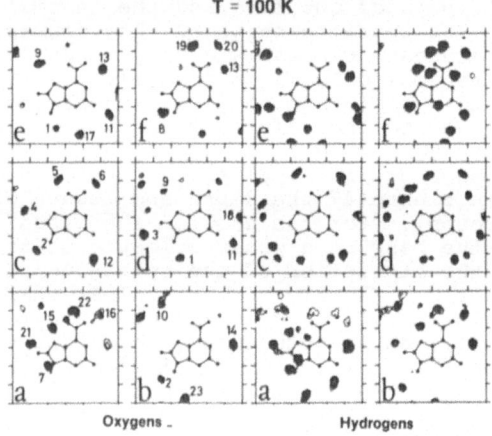

Figure 54. Probability density maps from M.C. simulations (adenine).

N(1), N(2) and N(3), attract one water molecule located below the plane, W(14), W(12) and W(15), respectively. In the plane the situation is less clear, because of the strong perturbation of nearby hydrophilic groups; however, W(4) can be assigned to N(4), even if this molecule is displaced from its ideal position by W(5). Above the plane, W(17) can be attributed to N(2) and W(9) to N(4). Two water molecules W(3) and W(11) are definitely part of the first solvation shell of adenine, and constitute (essentially) bridges between W(2) and W(4) and between W(1) and W(6), respectively. The shape of the probability distribution is particularly interesting. For example, W(5) and W(6) can translate by a considerable amount (up to 2 Å). Because of the sharp water-water replusion, if one water molecules moves, the second will also move to keep the same water-water separation.

Therefore, the two motions are expected to be concerted. In turn, these motions cannot be independent from those of the second solvation shell.

From the probability distribution we tentatively suggest that the water molecules in the plane might have a tendency to rotate around the periphery of the solute. The most probable positions are those reported. Since the total angular momentum in our system is nearly constant, a counter-rotation response might be postulated, for example by water molecules in the second hydration shell. Molecular dynamics computations and relaxation experiments could be very appropriate to confirm this hypothesis.

The temperature effect is analyzed by comparing the probability distribution maps obtained at T=300 K with those obtained at T=100 K (see Figure 54). The increased localization of the water molecules is very evident and needs no additional comment. In addition, by decreasing the temperature, and therefore the thermal motion, more molecules can be compacted into the same volume. For example, the $-NH_2$ group at T=100 K is solvate by more molecules than at T=300 K. A detailed comparison is made in Table 46, where for each water molecule the main hydrated group is given.

Table 46. Comparison of hydration at T=100K; ((a) is the water identification number, (b) the main hydration site)

T=300 K				T=100 K			
(a)	(b)	(a)	(b)	(a)	(b)	(a)	(b)
1	N(3)-H	2	N(3)-H	1	N(3)-H	2	N(3)-H
3	bridge	4	N(4)	3	bridge	4	bridge
5	NH_2	6	NH_2	5	NH_2	6	NH_2
7	N(3)-H	8	N(3)-H	7	N(3)-H	8	N(3)-H
9	N(4)	10	NH_2	9	N(4)	10	bridge
11	bridge	12	N(2)	11	bridge	12	N(2)
13	NH_2	14	N(4)	13	NH_2	14	N(1)
15	N(4)	16	NH_2	15	N(4)	16	NH_2
17	N(2)			17	N(2)	18	N(1)
				19	NH_2	20	NH_2
				21	bridge	22	NH_2
				23	NH_2		

At T=300 K, sixteen water molecules have a well-defined probability in the first hydration shell of cytosine, as seen in Figure 55. The C=O group is hydrated by three water molecules, one in the plane, W(1), one above, W(8) and one below, W(11). The $-NH_2$ group is hydrated by six water molecules, two in the plane, W(3) and W(4), two above, W(9) and W(10) and two below, W(13) and W(14). The N(2)-H group is hydrated by four molecules W(6), W(7) in the plane, and W(15), W(16) below the plane. The N(3) atom is solvated primarily by W(2). The two water molecules W(5) and W(12) are essentially bridges between the water molecules W(4) and W(15).

At T=300 K, eighteen water molecules have a well defined probability in the first hydration shell of guanine, as shown in Figure 55. The C=O group is solvated in the plane by W(7) and W(8), above the plane by W(14) and below by W(11). The $-NH_2$ group is solvated in the plane by W(1), W(13) and probably by W(2), below the plane by W(16). A water molecule above the plane, W(15), that solvates mainly N(1)-H is strongly perturbed by the $-NH_2$ group. The nitrogen atoms N(3) and N(5) are solvated by W(4), and W(18), respectively and perturb significantly the water molecules W(7) and W(16). The W(1)-H group is solvated by W(15); the cluster of water molecules W(1), W(8), W(11) and W(14) is strongly perturbed by the field of N(1)-H, but can be considered to belong mostly to the solvated of $-NH_2$ and C=O.

Thymine is solvated by a cluster of water molecules, eighteen of which have a very well-defined probability as indicated in Figure 55. The C=O(1) group is solvated by W(2) and W(4) in the plane, by W(3) above the plane, and W(15) below the plane. The C=O(2) group is solvated by W(18) in the plane, W(7) above the plane, and W(11) below the plane. The N(1)-H group is solvated by W(5) and W(8) above the plane, by W(6) and W(12) below the plane. The N(2)-H group is solvated by W(9), W(10) and probably W(17) above the plane and by W(13), W(14) and W(16) below the plane. One water molecule W(1) can be considered as bridging W(16) and W(17).

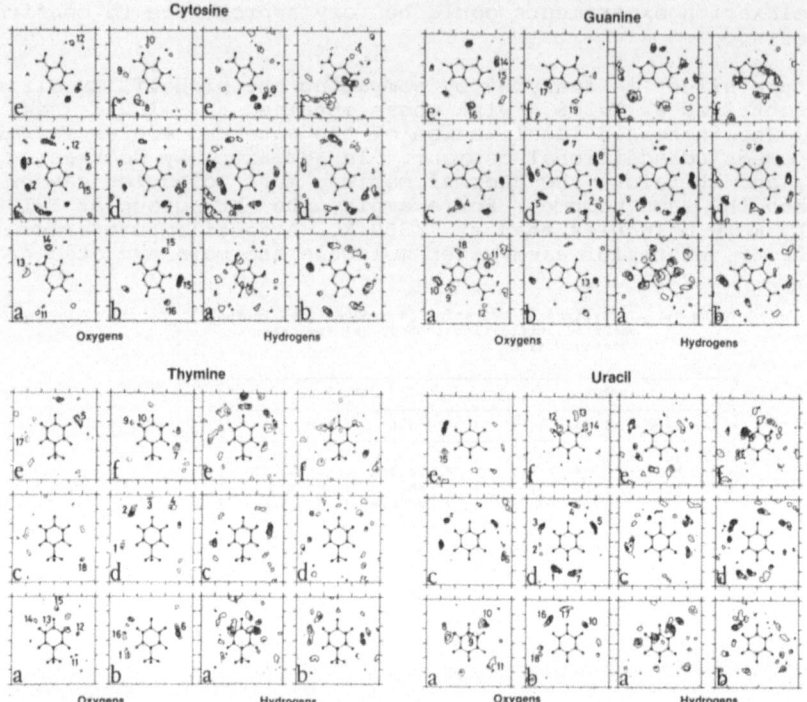

Figure 55. Probability density maps from M.C. simulations for other bases.

Eighteen water molecules can be identified in the first hydration shell of uracil as shown in Figure 55. Four water molecules hydrate C=O(1), namely W(4) in the plane, W(13) above, W(17) and W(9) below the plane, respectively. Three water molecules hydrate C=O(2), namely W(7) and W(6) in the plane, W(11) below the plane. The N(1)-H and N(2)-H groups are solvated by W(5), W(14), W(10) and W(3), W(8), W(12), respectively; W(16) is likely to be part of the hydration cluster for N(2)-H, but is strongly perturbed by the C=O(1) group. Two water molecules W(1) and W(2) make a double bridge between W(3) and W(7); one water molecule, W(18) reinforces this bridge from below the plane. The shape of the probability volume for the molecules in the plane is rather elongated, with expectation of the dynamical effects as discussed for adenine.

The five bases considered here are composed of hydrophobic groups such as C-H and CH_3 and hydrophilic groups, such as C=O, $-NH_2$, N-H and the atom N with a lone pair of electrons. The ability to attract one or more water molecules at a given position depends on the overall perturbation by the neighboring groups. However, some conclusion (of statistical type) can be tentatively advanced. In Table 47 we summarize the previously reported identification for the water molecules (to save space we write simply n to indicate W(n)); the water molecules in the plane are given first, followed by those above the plane and then by those below the plane (a semicolon separates water molecules in different planes).

The $-NH_2$ is solvated, in general, by a cluster of at least four water molecules, two in the plane, one above and one below; the C=O group is solvated by about three water molecules one in the plane, one above and

Table 47. Water molecules hydrating a given group

Base	Group	Water Molecules	Base	Group	Water Molecules
C	CO(1)	1;8;11	A	NH_2	5,6;13;10,16
G	CO(1)	7,8;14;11	C	NH_2	3,4;9,10;13,14
T	CO(1)	2,4;3;15	G	NH_2	1,2,3;16;12,13
T	CO(2)	18;7;11			
U	CO(1)	4;13;9,17			
U	CO(2)	7;6;11	A	N(3)-H	1,2;8;7
			C	N(2)-H	6,7;15,16;
A	N(1)	; ;14	G	N(1)-H	;15;
A	N(2)	;17;12	G	N(4)-H	5,6;17;9,10
A	N(4)	4;9;15	T	N(1)-H	;5,8;6,12
C	N(3)	2; ;	T	N(2)-H	;9,10,17;14,16
G	N(3)	4; ;	U	N(1)-H	5;14;10
G	N(5)	; ;18	U	N(2)-H	3;12;8,16

one below. The NH group has the same distribution, but seems to be more affected by the perturbation of neighboring groups. The nitrogen with a lone pair is solvated by one water molecule at least; the notion that this water molecule in solution must be in the plane (see also the discussion below) finds little support from our simulation.

The average water-water interaction energy, $\bar{U}(w,w)$ computed in the 400,000 configurations (after equilibration) for the forty water molecules is given in Table 48. The interaction energy of the water molecules with a base $\bar{U}(w,b)$ is also given, as an average value, in Table 48.

The water-water interaction in bulk water, $\bar{U}°(w,w)$, using the same potentials is -8.51 ± 0.14 Kcal/mole, at T=300K. Therefore, a system of forty water molecules and a base compared with a system of forty water molecules in bulk water, gains the $\bar{U}(w,b)$ energy, but loses an energy amount equal to $40\{\bar{U}°(w,w)-\bar{U}(w,w)\}$; the net balance is therefore $\Delta\bar{U}=\bar{U}(w,b)-40\{\bar{U}°(w,w)-\bar{U}(w,w)\}$ reported in Table 48.

Table 48. $\bar{U}(w,w)$ and $\bar{U}(w,b)$ in Kcal/mole.

Base	$\bar{U}(w,w)$	$\bar{U}(w,b)$	$\Delta\bar{U}$	T(K)
A	-6.58 ± 0.04	-64.09 ± 0.36	13.0	300
A	-8.01 ± 0.03	-73.35 ± 0.43	-53.3	100
G	-5.69 ± 0.06	-154.33 ± 1.17	-41.5	300
C	-6.41 ± 0.05	-68.30 ± 0.79	15.7	300
T	-6.44 ± 0.06	-36.99 ± 0.91	45.8	300
U	-6.43 ± 0.04	-77.18 ± 0.55	2.0	300

We could now deduce the overall hydration preference of the bases in DNA and RNA. Considering only $\bar{U}(w,b)$ we obtain the following ordered sequence:

for DNA G(-154)>C(-68)≈A(-64)>T(-39)
for RNA G(-154)>U(-77)>C(-68)≈A(-64)

Considering $\Delta\bar{U}$, we obtain the following ordered sequence:

for DNA G(-42)>A(13)≈C(16)>T(46)
for RNA G(-42)>U(2)>A(13)≈C(16).

The two sequences are essentially equal, allowing for an error of ± 4 Kcal/mole due to the statistical deviations reported in Table 48 and to the limitation of the method. These extrapolations are common in the nucleic acids literature. In our opinion, however, it constitutes mainly a rationalization on available experimental data; the neglect of field effects deprive the extrapolation of physical validity.

As a final comment, we add the computed interaction energy between one water molecule and a single base obtained by using pair-potential (80) for few positions and orientations, namely those reported in Figure 47. Such interaction energies are given in Table 49, where for each base the numerals refer to a water molecule of Figure 47 followed by the interaction in Kcal/mole. Let us now consider the base-pairs.

Table 49. Water-base interaction energy (in Kcal/mole) from analytical potentials for the water molecules reported in Figure 47.

A		C		G		T		U	
1	−9.94	1	−7.70	1	−13.24	1	−6.18	1	−5.50
2	−8.90	2	−6.07	2	−5.92	2	−10.65	2	−7.65
3	−6.32	3	−9.02	3	−6.85	3	−6.10	3	−9.61
4	−4.18	4	−8.52	4	−10.61	4	−9.99	4	−7.11
5	−8.15	5	−7.21	5	−8.03	18	−5.79	5	−10.64
6	−7.72	6	−9.02	6	−5.30			6	−6.83
11	−4.45	7	−12.14	7	−16.59			7	−6.58
				8	−14.20				

In Figures 56-58 only water molecules with high probability are identified. Whenever a water molecule has a PD that extends in a region connecting more than one volume, the identification is given for only one of the volumes. The primed and the unprimed code numbers correspond to those previously used to identify water molecules in the two separated bases of the base-pair in order to simplify the comparison between hydration in the separated bases and in the base-pairs (see discussion for Figure 47).

The PDs' maps for oxygen and hydrogen atoms of A-T are reported for the six volumes in Figure 56. Five water molecules, W(1),......,W(5) hydrate the A-base of A-T in the molecular plane at about the same position and orientation as found previously for A; similarly one water molecule, W(1'), hydrates T. Water molecules outside the molecular plane and determined previously for the separated bases are W(8), W(9'), and W(17') above the plane and W(7), W(15), W(15'), W(14') and W(16') below the plane. Two water molecules W(1") and W(2") form hydrogen-bond bridges: W(1") with the CH atoms of A and the CO group of T, W(2") with the NH_2 and the CO groups of A and T, respectively. A few additional hydration sites with rather low probability, such as W(3") and W(4"), can be noticed in the figures; in total we have identified eighteen water molecules in the first hydration shell of the A-T pair.

The water-water interaction energy, averaged over the fifty water molecules and 500,000 configurations, $\bar{U}(w,w)$, is −6.27±0.07 Kcal/mole; the total solute-solvent interaction energy, average over the 400,000 configurations, $\bar{U}(w,p)$, is −65.29±1.24 Kcal/mole. The latter value is smaller than the sum of the solute-solvent interactions, previously obtained for the separated bases, by about 36 Kcal/mole. This is in agreement with the finding that the total number of water molecules in the first hydration shell is smaller than the sum of the water molecules in the first shell of the separated bases, A and T, reported as seventeen and eighteen, respectively.

Figure 57 reports the PDs for oxygen and for hydrogen atoms of the A-U base-pair. Nine water molecules in the molecular plane correspond approximately to the position and orientation previously obtained from the individual bases, namely the water molecules W(1), W(2), W(3), W(4), W(5), W(11), W(1'), W(2') and W(3'). Outside the molecular plane five additional water molecules W(7), W(8), W(10), W(12) and W(8') can be similarly characterized. Three molecules appear to form hydrogen-bonds with those atoms bridging the two bases, namely W(1") in the plane,

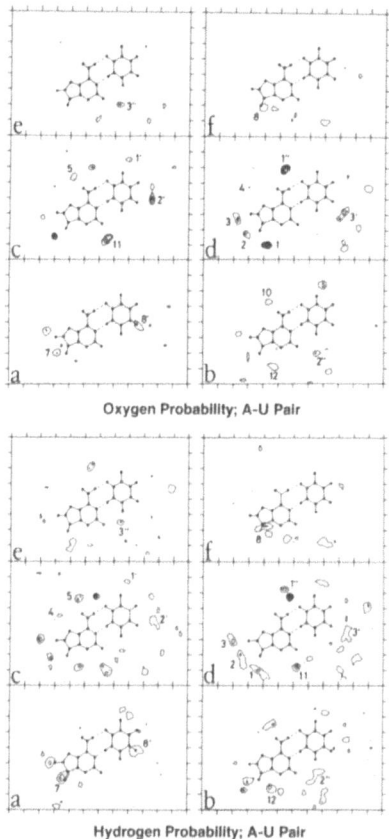

Figure 56. Probability density maps from M.C. simulations for A-U base-pair.

W(2") below the plane and W(3") above the plane; two water molecules W(2") and W(3") interact via hydrogen-bond. The total number of hydration sites in the first shell is about seventeen; evidences of a second solvation shell is rather clearly visible in the PDs of Figure 57.

The computed value for $\bar{U}(w,w)$ is -6.8 ± 0.05 Kcal/mole; the solute-solvent interaction is $\bar{U}(w,p) = -98.61 \pm 2.58$ Kcal/mole, about 43 Kcal/mole less than the sum of the solute-solvent interaction for the two bases.

In Figure 58 the PDs for the G-C pair are reported. We easily recognize twenty water molecules in the first hydration shell; nine approximately correspond to hydration sites of G and five to hydration sites of C, previously identified in the solvation study of the individual bases. Six new water sites are obtained by analyses of Figure 58: W(1") hydrogen-bonded to the CO of G and to W(4'); W(2") located above the molecular plane and hydrogen-bonded mainly to the -NH...N hydrogen-bond of the base pair; W(3"), with the same role as W(2"), but located below the molecular plane; W(4") interacting with the NH_2 group of C and below the plane; W(5") interacting with the NH_2 group

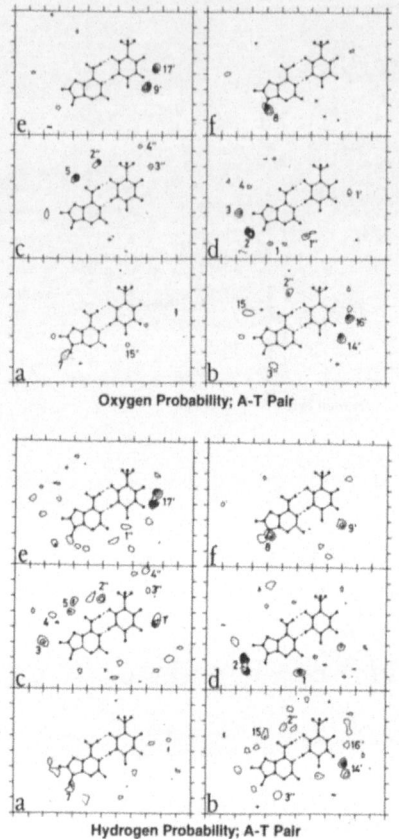

Figure 57. Probability density maps from M.C. simulations for A-T base-pair.

of G and with W(3) and W(1'); finally W(6") interacting with W(1"), W(7) and the CO group of G. The value of $\bar{U}(w,w)$ is computed as -5.50+0.06 Kcal/mole, smaller than the value either for G (-5.69 Kcal/mole) or for C (-6.41+0.05 Kcal/mole). The $\bar{U}(w,p)$ is -198.18+1.10 Kcal/mole, less than the sum of the corresponding quantity for the isolated bases in solution by about 24 Kcal/mole.

This difference is the smallest one found for the three base-pairs analyzed here. Let us now discuss the base-pair solvation in a more general way.

The number of water molecules solvating the base-pair is expected to be smaller than the sum of the numbers of water molecules solvating the two bases of the pair, taken separately, since part of the periphery of one base is connected to part of the periphery of the second base. This qualitative expectation is quantitatively verified in this work. In terms of solvent-solute interactions, this expectation corresponds to $\bar{U}(w,p)$ being smaller, in absolute value, than the sum of two separate

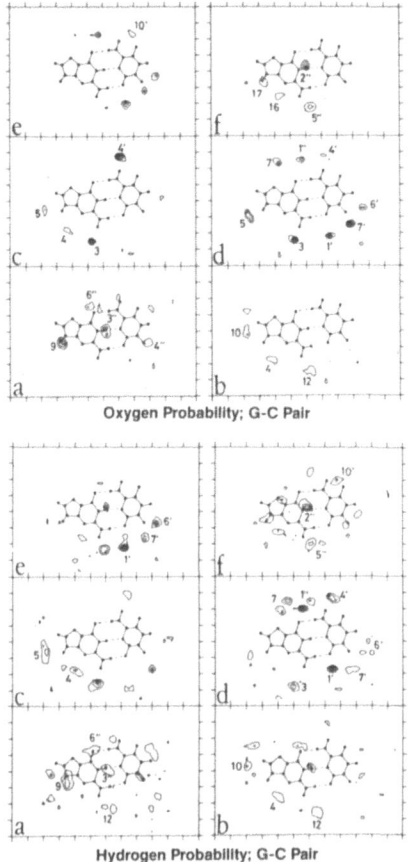

Figure 58. Probability density maps from M.C. simulations for G-C base-pair.

$\bar{U}(w,b)$ interaction energies. This decrease is 24 Kcal/mole for G-C, 36 Kcal/mole for A-T and 43 Kcal/mole for A-U.

All bases pairs present new features relative to the hydration obtained for the individual bases, the main one being four water molecules hydrating the base-base hydrogen-bonded region in G-C, and two water molecules hydrating the equivalent region in A-T and in A-U. This "new" feature is probably more apparent than real, since, for example, in the PDs for guanine there are indications of hydrogen atoms below the molecular plane (see Figure 55, inserts a and f) corresponding very roughly to the water molecules W(2") and W(3").

The old problem related to the proton tunneling in the hydrogen-bond for the base-pair that has found limited support in preliminary computations on a base-pair in vacuo (187), can now be restated; not only the motion of one proton is likely to be concerted with the motions of protons in the neighboring hydrogen bonds of the base-base system, but also the external field of the solvating water molecules (if we consider base-pairs in solution) or the neighboring bases (if we consider a base-pair in DNA or RNA) must be included in any realistic simulation of the proton tunneling rate constants.

5.12 SOLVATION OF B-DNA DOUBLE HELIX AT T=300°K

Previously we have reported on the interaction between one water molecule and the bases (80) and base-pairs (194) of the nucleic acids, A-DNA single helix (189), B-DNA single (192) and double helix (190, and 192). In addition, Monte Carlo simulations (at 300°K) have been presented for a cluster of water molecules enclosing the bases and the base-pairs (194), or a limited region around A-DNA single helix (189) and B-DNA single helix (190 and 192). These studies represents preliminary steps. We extend our previous effort by considering, via simulations, not only a much larger number of water molecules than previously but also the effect of counter ions, initially the Na^+ ion; we report on qualitative features of Na^+-B-DNA at 300°K double helix and on quantitative aspects of B-DNA in solution at 300°K. In a following study (see Section 5.13) we shall present a quantitative study of Na^+-B-DNA in solution. The B-DNA double helix fragment we consider has been previously discussed (see Section 5.11 and Figure 53, in particular), and consist of twelve base-pairs (namely, two more base-pairs than needed to reproduce a full B-DNA double helix turn) with the corresponding sugar and PO_4^--CH_2 units. The B-DNA double helix fragment is enclosed into a cylinder with its axis co-axial to the B-DNA long axis (z axis). The cylinder height is 36.0 Å with a base radius of 14.5 Å. The two base-pairs and the corresponding sugar units kept outside the cylinder (one above and one below) have been added in order to improve the interaction field descriptions at the bottom and top ends of the B-DNA fragment. In the Monte Carlo simulation below reported 447 water molecules have been placed into the cylinder. The equilibration process was carried out for 2×10^6 conformations; the statistical data below analyzed are obtained from additional 2×10^6 Monte Carlo "moves" (these computations have been carried out on an IBM 370/3033 computer).

Some of the analysis below reported is not carried on with using probability density maps, as done in the past (189, 190, 192, 194; see also 76, 161 and 163) since these are somewhat difficult to read and even more difficult to use as input data. The probability maps have been replaced by a model based on a new algorithm described in the following four steps:

1. after computation of the probability density maps, the probability density maxima for the oxygen atoms are located (neglecting low probability maxima by selecting a threshold that ensures to limit the number of maxima to 447, the number of water molecules enclosed into the cylinder),

2. for the hydrogen atoms, the probability maxima are determined subject to the constraint of being located on a sphere of radius equal to the O-H internuclear separation in H_2O,

3. a sphere of radius 0.5 Å is centered at the oxygen and at the two hydrogen atoms probability maxima, and

4. the 2×10^6 conformations of the Monte Carlo simulation are scanned to determine how many times a water molecule fell into the volume defined in 3.

We note that in the probability density maps, the distance between the probability maxima of an oxygen atom and its associated hydrogen atoms is nearly, but not necessarily, exactly equal to the O-H distance in H_2O; thus, in our newly proposed algorithm, we loose some of the information available from the probability maps. In addition the assumption of a sphere around the oxygen and hydrogen atoms, implies an isotropic probability distribution; as pointed out previously (189, 190,

192) the probability distribution is often anisotropic, especially at room temperature. On the other hand, the advantage of the new method is that it allows to obtain a graphical representation of immediate understanding and replaces the snap-shot pictures often used in the analysis of Monte Carlo data. As known, such pictures are limited to only one conformation, and therefore, have no statistical value. The technique here described, brings 50% to 70% of the full set of the simulated conformations into the three spherical volumes associated to each water molecule. By increasing the sphere's radii to 1A, 90% to 95% of the computed configurations are accounted for; however, this larger radius decreases the information content concerning the relative orientation of the molecules in the solvent.

Let us start with a gross analysis of the computed data. In Figures 59, 60 and 61 we report the water molecules solvating either the phosphate groups in each of the two helix of B-DNA or the water molecules contained in a disk of about 5A thickness, (slicing the cylinder perpendicular to the z-axis). The phosphate, the sugar groups and the bases connected to one of the two helix are designated with an asterisk in order to differentiate it from the equivalent groups of the second helix; equivalently the two helix are referred as h and h*, for short. In the figures, the water molecules are represented with the new algorithm; we shall talk of "water molecule number N", as a short expression to indicate "the ensemble of water molecules that falls within the volume number N, consisting of the previously described three spheres of radius 0.5A."

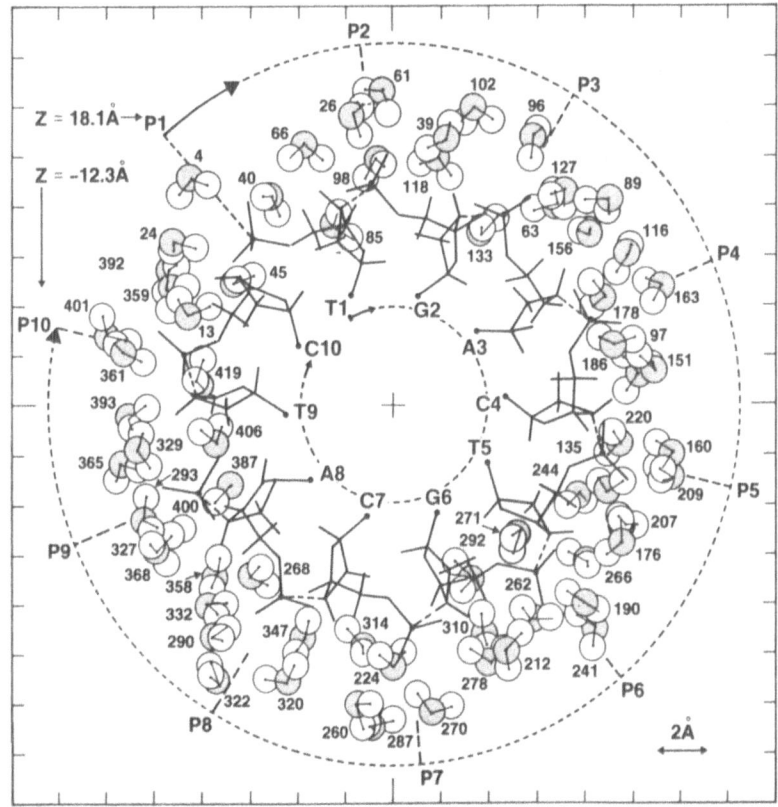

Figure 59. Water molecules solvating the h helix in B-DNA.

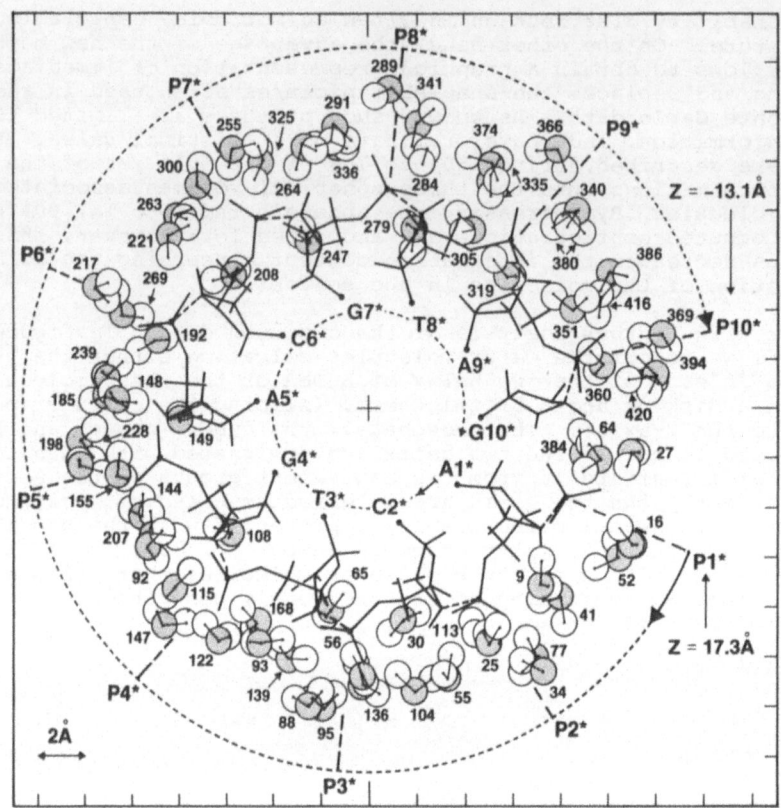

Figure 60. Water molecules solvating the h* helix in B-DNA.

In Figure 59, we report the ten phosphate groups P1 to P10 of h and the corresponding sugar unit, but not the base-pair, that are, however, indicated by reporting the terminal nitrogen atom and by using the notation, T1, G2, ..., C10. The outmost circumference has 14.5A radius; the marks on the figure's frame are at 2A interval. The water molecules are seen from positive values looking down toward negative z values. Only the water molecules very near the phosphate P1 (at z = 18.1A) to P10 (at z = -12.3A) are reported. Since the "water molecules" are numbered with an index of increasing value along the z direction (from positive z to negative z), low indices (starting from 1) corresponds to water molecules solvating the top of the B-DNA fragment, high indices (approaching 447) correspond to water molecules solvating the bottom of the B-DNA fragment.

Notice in Figure 59 the dual features of the water clusters: not only the water molecules enclose the PO_4^- group, but also form hydrogen bonded filaments (see for example, the water molecules 322, 230, 332, and 358 in the P8-P9 region). The data in Figure 60 provides a view of the water molecules solvating the phosphate groups of the h* helix. The water molecules in this figure seems to spread somewhat less along radial lines and to be confined on circular patterns. Close analysis

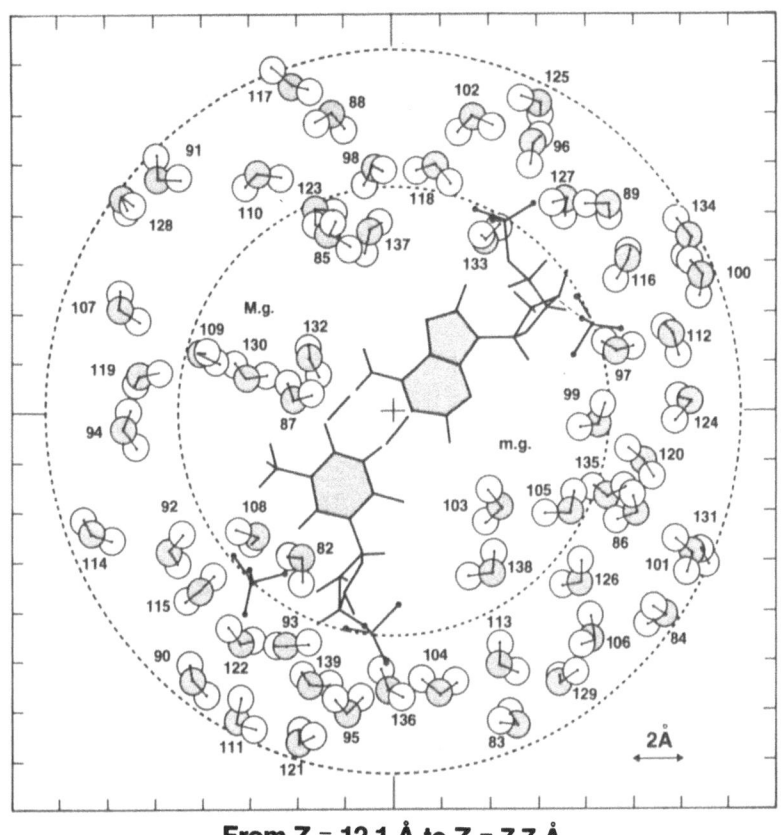

From Z = 12.1 Å to Z = 7.7 Å

Figure 61. Water molecules contained in a disk of 4.4 A thickness and solvating B-DNA; the radii of the two circumferences are 14.5A and 8.8A, respectively.

(provided by the data in the tables that will follow) sharply mitigate (but do not fully eliminate) this first hand impression due to, a) to sterical visual effects, and b) an insufficient number of water molecules reported in the two figures. In Figure 61, we report only one base-pair, the A3-T3* base-pair; the water molecules experience the immediate fields of the G2-C2* and C4-G4* base-pairs (not reported in the Figure in order not to further complicate the drawing).

Notice, in addition that in the major groove (M.g.) there are ten water molecules (82, 108, 87, 132, 130, 109, 123, 85, 137 and 133) whereas only four water molecules are found in the minor groove (138, 103, 105, and 99); this remark changes to its opposite, when we extend the major groove and the minor groove volume up to a radius of 14.5A. The physical reason for these findings is that near to the bases there is more "free" space in M.g. than in m.g., but further out toward R = 13 or 14A the field generated by the phosphates in the m.g. is stronger than the field generated by the phosphates in the M.g.

Table 50. Water Molecules Solvating The Phosphates In The h Helix.

P#	W#	ATOM(1)	R(H1)	ATOM(2)	R(H2)	E(W-DNA)	E(W-W)
P1	4	O1P	1.7	O1P	2.8	-115.3±6.1	1.8±3.5
	13	O2P	2.7	O2P	2.1	- 86.9±6.5	1.5±1.2
	24	O2P	3.0	O2P	1.8	-114.5±5.0	0.5±3.8
	40	O1P	1.9	O5'	3.1	- 90.0±6.2	-12.0±4.0
	45	O2P	1.5	O2P	2.9	- 92.7±3.7	- 9.3±4.2
P2	26	O1P	2.8	O3'	2.1	-103.9±4.2	4.6±4.5
	39	O1P	2.9	O1P	1.7	-104.8±5.7	- 2.2±1.6
	61	O1P	1.6	O1P	2.9	-121.2±5.7	3.9±3.7
	66	O2P	1.7	O2P	2.8	-113.9±4.8	0.8±4.5
	68	O2P	2.6	O2P	3.0	- 59.3±3.8	-17.4±4.0
	70	O1P	4.2	O1P	3.6	- 38.0±7.5	-25.2±5.3
	85	O2P	1.6	O2P	2.9	-104.1±3.5	7.5±3.8
	98	O1P	2.4	O2P	3.0	- 89.3±3.9	8.8±4.8
P3	63	O3'	1.9	O3'	3.2	- 60.4±8.9	-18.3±1.6
	89	O1P	2.4	O1P	3.2	- 99.0±5.8	4.0±3.5
	96	O1P	1.6	O1P	2.7	-113.7±4.8	-10.4±3.8
	102	O2P	3.2	O2P	3.4	- 78.3±3.2	-11.6±2.4
	118	O2P	1.6	O2P	2.9	-117.7±5.6	-12.8±3.3
	127	O1P	1.7	O1P	3.1	-124.5±4.8	7.6±3.5
	133	O2P	1.8	O2P	3.0	- 70.6±5.9	-21.7±3.5
P4	97	O3'	1.8	O3'	2.9	- 75.2±4.6	-22.9±2.6
	112	O1P	2.2	O1P	3.2	- 81.5±9.2	-15.2±3.4
	116	O1P	3.3	O3'	2.1	- 87.8±6.6	0.3±3.5
	151	O1P	1.7	O1P	2.8	-110.3±6.7	- 1.0±4.2
	156	O2P	1.5	O1P	2.8	-119.7±4.3	-10.8±4.3
	163	O1P	1.7	O1P	2.8	-125.1±6.8	2.3±4.5
	178	O2P	2.1	O2P	3.0	- 82.4±3.6	- 9.5±3.5
P5	135	O3'	1.8	O3'	2.9	- 61.9±3.8	-18.2±4.3
	160	O1P	2.0	O1P	2.9	-104.2±5.0	2.0±5.3
	176	O1P	1.7	O1P	3.1	-117.9±7.2	2.9±4.5
	186	O2P	1.6	O2P	2.8	-121.5±3.6	-13.7±3.2
	209	O1P	1.7	O1P	2.8	-130.8±4.8	0.2±3.5
	223	O2P	3.6	O2P	3.8	- 44.5±7.3	-24.8±4.2
P6	190	O3'	1.8	O3'	3.2	- 80.3±5.9	-12.2±4.7
	212	O1P	1.6	O1P	2.9	- 85.4±7.5	-10.5±4.0
	241	O1P	1.6	O1P	2.8	-105.8±5.6	- 8.9±4.3
	244	O2P	1.6	O2P	2.7	- 91.1±3.5	-10.0±4.0
	262	O1P	2.0	O1P	2.8	- 77.3±7.2	-14.6±3.0
	266	O2P	2.1	O2P	2.6	-114.3±6.1	2.1±5.4
	271	O2P	2.3	O2P	3.5	- 75.7±7.0	-18.0±5.6

Table 50. Continued

P#	W#	ATOM(1)	R(H1)	ATOM(2)	R(H2)	E(W-DNA)	E(W-W)
P7	218	O3'	3.9	O3'	4.3	- 61.3±2.2	-23.4±4.0
	224	O3'	1.7	O3'	3.1	- 95.3±4.3	- 6.9±2.1
	230	O3'	3.5	O3'	3.8	- 49.7±7.9	-26.2±5.5
	260	O1P	1.9	O1P	2.7	-105.0±6.2	0.1±5.0
	270	O1P	1.7	O1P	2.9	-125.6±4.5	7.1±5.1
	278	O1P	1.7	O2P	2.9	-128.3±4.6	- 4.8±3.4
	287	O1P	3.0	O1P	3.0	- 52.3±9.7	-19.8±4.7
	292	O1P	1.6	O2P	2.9	-104.9±5.1	2.1±5.0
	310	O2P	3.0	O2P	3.1	- 73.5±7.6	-14.7±5.6
P8	268	O3'	1.8	O3'	3.1	- 90.3±5.2	-13.5±3.9
	290	O1P	1.6	O1P	3.0	-121.2±5.8	2.9±3.2
	314	O2P	1.7	O2P	3.0	-113.5±11.2	3.0±2.9
	320	O2P	2.5	O1P	2.8	- 93.1±9.8	- 9.2±3.6
	322	O1P	2.9	O1P	2.9	- 54.0±10.7	-17.8±3.9
	332	O1P	1.7	O1P	2.9	-108.5±5.4	-10.6±2.9
	347	O2P	1.8	O2P	2.7	-114.0±3.4	-11.3±4.2
P9	293	O3'	3.7	O3'	3.9	- 54.2±3.8	-13.0±2.7
	327	O3'	2.0	O3'	3.1	- 76.3±6.6	- 9.1±5.1
	329	O1P	1.9	O1P	2.8	- 93.2±3.4	1.0±3.9
	358	O2P	1.5	O2P	2.8	-125.3±4.7	5.7±4.2
	365	O1P	1.6	O1P	2.8	- 94.7±5.6	7.4±3.2
	368	O1P	1.9	O2P	2.7	-119.3±10.6	8.5±4.4
	387	O1P	1.8	O2P	3.3	-109.2±7.4	- 4.1±3.1
P10	359	O3'	2.4	O3'	3.5	- 86.6±4.1	- 5.7±3.4
	361	O3'	1.9	O1P	2.5	- 85.3±6.9	8.4±4.2
	379	O1P	3.3	O1P	3.9	- 41.5±6.6	-17.2±4.3
	392	O1P	1.6	O1P	2.8	-109.7±5.0	- 5.3±2.8
	393	O1P	1.7	O2P	3.0	-124.3±4.5	3.1±4.7
	401	O1P	1.6	O2P	2.6	-116.7±3.0	- 5.5±3.8
	406	O2P	2.0	O2P	2.8	- 83.8±7.9	- 3.2±5.6
	419	O2P	1.7	O2P	3.1	-109.4±7.0	- 0.8±2.6
P11	381	O3'	3.1	O3'	4.0	- 53.8±3.5	-24.3±3.9
	408	O1P	1.6	O1P	2.9	-113.0±5.1	- 5.7±5.0
	412	O1P	3.1	O3'	3.5	- 63.4±6.7	-16.9±2.8
	422	O1P	3.1	O1P	3.3	- 68.8±4.7	- 1.0±5.6
	432	O2P	1.8	O2P	2.8	- 79.3±4.9	-18.0±5.6
	433	O1P	1.6	O1P	2.7	-118.3±4.3	8.5±2.5
	439	O2P	1.8	O2P	2.8	-120.4±4.3	2.5±4.4
AV	(58)		1.75		2.88	103.4±5.5	- 4.5±3.8

Table 51. Water Molecules Solvating The Phosphates In The h* Helix.

P*#	W#	ATOM(1)	R(H1)	ATOM(2)	R(H2)	E(W-DNA)	E(W-W)
P1*	9	O2P	1.6	O2P	2.8	-116.2±5.5	- 5.2±3.5
	16	O1P	1.7	O1P	2.8	-114.9±4.1	- 2.8±3.1
	27	O1P	1.6	O1P	2.8	-114.3±3.9	- 3.5±4.0
	52	O3'	3.2	O3'	3.3	- 58.8±4.2	- 16.6±4.7
	64	O3'	2.2	O3'	3.6	- 76.1±4.1	- 16.6±3.7
P2*	25	O1P	3.1	O2P	2.0	-109.7±5.3	- 3.5±4.4
	30	O2P	1.7	O2P	2.9	-115.3±6.7	3.2±1.6
	34	O1P	2.7	O1P	2.8	- 77.8±6.1	- 13.8±3.5
	41	O1P	1.7	O2P	2.6	- 83.9±4.8	- 16.4±4.2
	55	O1P	2.0	O1P	2.6	- 86.4±6.0	- 7.3±3.1
	77	O1P	1.6	O1P	2.9	-104.6±6.5	- 9.4±3.9
	113	O3'	3.0	O3'	4.3	- 27.1±5.1	- 30.3±4.0
P3*	56	O1P	2.8	O2P	3.3	- 82.3±9.5	- 7.5±3.8
	65	O2P	1.7	O2P	3.0	-114.5±4.2	- 3.0±2.6
	75	O1P	3.5	O1P	4.1	- 64.1±4.3	- 12.6±3.8
	82	O2P	2.6	O2P	3.6	- 62.8±7.8	- 21.2±3.5
	93	O2P	1.7	O2P	3.0	-120.2±6.7	- 7.8±3.5
	95	O1P	1.7	O1P	2.9	-120.7±6.0	- 1.5±5.6
	104	O1P	1.6	O1P	3.0	-114.8±6.7	- 2.9±4.7
	136	O3'	2.7	O3'	1.9	- 85.6±5.2	- 7.5±4.6
	139	O3'	3.0	O3'	3.8	- 48.7±8.8	- 19.8±3.1
P4*	108	O2P	1.7	O2P	2.8	- 80.4±6.0	- 14.3±5.2
	115	O2P	1.8	O2P	2.8	-112.3±6.7	- 4.6±3.6
	122	O1P	1.7	O1P	2.4	-106.2±7.4	- 7.8±3.5
	144	O2P	1.7	O2P	3.1	-111.4±8.5	- 8.4±2.5
	147	O1P	1.7	O1P	3.0	-119.4±7.2	- 6.9±4.5
	168	O1P	1.9	O1P	3.4	- 74.5±13.2	- 11.2±3.3
P5*	148	O2P	2.8	O2P	3.0	- 84.5±6.3	- 10.6±5.4
	149	O2P	1.7	O2P	3.1	-115.3±7.1	0.0±2.9
	155	O1P	1.8	O2P	2.9	-121.1±8.2	1.3±5.0
	180	O1P	3.3	O1P	3.5	- 71.8±4.9	- 7.6±5.1
	185	O2P	1.7	O2P	2.9	-106.4±6.9	- 12.5±3.0
	198	O1P	1.6	O1P	3.0	-120.5±6.8	0.8±3.6
	207	O1P	1.6	O1P	2.8	- 92.3±4.8	- 5.9±4.2
	228	O3'	3.4	O3'	3.7	- 57.9±6.8	- 24.3±2.7
	246	O3'	2.9	O3'	3.7	- 65.3±1.1	- 13.8±5.6
P6*	192	O1P	3.0	O5'	2.5	- 75.4±5.9	- 11.7±4.4
	208	O2P	1.8	O2P	2.5	- 94.3±7.0	- 14.8±3.3
	217	O1P	1.7	O1P	2.7	-130.7±4.0	1.8±3.4
	221	O2P	1.7	O2P	2.8	-127.5±4.2	- 1.2±3.6
	239	O1P	1.6	O2P	2.7	- 97.8±6.3	- 8.1±2.7
	269	O3'	1.8	O3'	3.0	- 84.2±4.9	- 19.9±4.5

Table 51. Continued

P*#	W#	ATOM(1)	R(H1)	ATOM(2)	R(H2)	E(W-DNA)	E(W-W)
P7*	237	O1P	2.6	O1P	3.3	- 54.9±4.8	- 22.6±5.1
	247	O1P	1.6	O2P	2.8	- 82.1±4.1	- 17.8±2.9
	250	O2P	3.3	O2P	3.7	- 67.3±3.3	- 11.2±3.8
	255	O1P	1.7	O1P	2.9	-127.9±5.8	1.0±4.7
	263	O1P	1.6	O1P	2.8	-118.4±6.9	8.1±4.4
	264	O2P	1.6	O2P	2.8	-119.2±4.3	- 7.9±3.7
	286	O1P	3.1	O1P	3.7	- 53.8±7.3	- 14.9±3.6
	300	O1P	1.8	O1P	2.7	- 95.4±4.6	- 5.5±4.2
	325	O3'	2.4	O3'	3.5	- 41.7±7.7	- 25.4±3.6
P8*	279	O2P	2.0	O2P	3.5	- 96.6±5.7	- 15.7±3.1
	284	O2P	1.7	O2P	2.8	-128.6±2.5	- 4.1±3.5
	289	O1P	3.3	O1P	3.3	- 71.1±7.8	- 7.0±3.6
	291	O1P	1.7	O1P	3.1	-126.7±7.9	0.1±5.4
	305	O2P	1.8	O2P	2.6	- 95.2±9.0	- 8.9±2.8
	326	O1P	2.8	O2P	3.0	- 56.1±10.6	- 19.2±3.6
	337	O1P	1.7	O1P	2.8	-118.0±4.5	- 3.1±2.6
	341	O1P	2.7	O2P	2.0	- 95.3±5.4	- 12.9±4.4
P9*	311	O2P	3.3	O2P	4.5	- 40.7±4.5	- 34.7±3.4
	319	O2P	1.7	O2P	3.1	- 89.9±5.3	- 17.3±3.1
	335	O1P	1.7	O1P	2.8	- 65.3±8.0	- 16.9±4.2
	340	O1P	2.1	O2P	3.0	-117.4±4.4	2.0±4.2
	351	O2P	1.6	O2P	2.7	- 88.0±5.2	- 17.8±2.5
	366	O1P	2.7	O1P	2.8	- 88.7±6.5	- 0.2±4.4
	374	O1P	1.6	O1P	3.1	-113.1±8.2	- 6.2±5.1
	380	O3'	1.9	O3'	3.1	- 90.9±4.7	- 8.3±3.7
	411	O3'	3.5	O3'	4.0	- 54.0±4.5	- 19.7±3.1
P10*	360	O2P	1.8	O2P	3.1	-103.5±5.0	- 10.6±3.3
	363	O1P	3.1	O1P	3.1	- 64.7±5.0	- 12.6±5.1
	364	O2P	3.0	O2P	3.2	- 76.0±5.3	- 17.6±3.9
	384	O1P	1.6	O1P	2.9	-106.6±5.3	- 9.8±4.2
	386	O1P	1.8	O1P	2.9	- 89.1±8.6	- 20.1±3.2
	394	O1P	1.8	O2P	2.9	-125.2±4.2	0.3±2.7
	416	O1P	2.2	O1P	2.4	-106.4±5.5	- 1.9±3.1
	420	O3'	2.6	O3'	3.5	- 75.9±4.3	- 2.6±3.1
P11*	399	O2P	1.7	O2P	3.1	- 93.5±6.7	- 3.3±5.0
	402	O1P	1.7	O1P	3.0	-115.6±4.0	- 5.6±3.2
	409	O2P	2.2	O2P	2.8	- 96.8±6.4	- 7.4±2.8
	421	O2P	1.7	O2P	2.8	-119.1±6.4	3.8±5.0
	435	O1P	2.1	O1P	2.6	-106.4±4.5	- 2.3±5.0
	441	O1P	1.8	O1P	2.6	-105.8±5.8	3.3±2.8
AV			1.74		2.86	-106.0±6.0	- 6.1±3.8

A more quantitative analysis is provided in Tables 50 and 51, where we report the water molecules solvating the phosphates of the h and of the h* helix, respectively. The first column reports the phosphate group identification; in the following columns we report the water index, the distance from one of the hydrogen atoms (H1) from either a free oxygen of PO_4^- (namely, O1P and O2P) or the bound oxygen atoms (O3' and O5'), the water-B-DNA average interaction energy in kj/mole, its mean standard deviation, and the water-water interaction energy in kj/mole and its standard deviation.

In Figures 62 to 65, we report the water molecules solvating the sugar unit and the base-pairs. One point should be immediately stressed: we report in these figures not only the water molecules hydrogen bonded to one (or more) bases but also some of those "nearby" the bases. For a given water molecule we use the following notation (in addition to the index for the water "volume"): strongly hydrogen bonded water molecules to B-DNA are differentiated from non-strongly hydrogen bonded one by writing a short identification of the solvated site either without or within a parentheses.

For example, in Figure 62 (Insert 1) the water molecules reported are in the vicinity of the T-A* and the G-C* base-pair; the base-pair (T-A*) is represented with full lines since at higher z-value than the lower base-pair (G-C*), represented by dashed lines. The water molecule 68 solvates strongly the second phosphate group in the h helix (P2); water 46 solvates strongly the zero-th sugar group in the h* helix (S0*); water 46 solvates strongly S0* and also adenine A1* of h* and (but less strongly) guanine G2 of h; water 49 is too far from any B-DNA atom to be assigned as solvating a specific group and therefore, it is labeled only as M.g., namely, one of the molecules in the major groove; finally water 73 is unlabeled, because neither strongly hydrogen bonded to any atom of B-DNA, nor within the major groove and, therefore, within the minor groove. In conclusion, we have considered three types of water molecules: strongly hydrogen bonded (to one or more atoms of B-DNA) weakly hydrogen bonded, very weakly hydrogen bonded. Somewhat arbitrarily, this classification is based on the value of the internuclear distance of the water's hydrogen (or oxygen) atoms from a given atom of the B-DNA fragment. More precisely, water reported as strongly hydrogen bonded to PO_4^- are those with an H--O hydrogen bond internuclear distance not larger than 2.1A. Water molecules with a hydrogen bond equal or shorter than 2.2A are considered as strongly hydrogen bonded either to a sugar or to a bases; if the hydrogen bond length is between 2.2 and 2.8A, then we classify it as a weak hydrogen bond. These criteria are rather restrictive and somewhat arbitrary: the PO_4^- field is very intense and a water molecule "strongly" hydrogen bonded to one of the oxygen atoms in PO_4^-, necessarily strongly feels the field of the remaining atoms in the PO_4^- group. For this reason we have not even attempted to list in the Tables those water molecules weakly hydrogen bonded to PO_4^-. In our classification of "strong hydrogen bond" we have included as an additional criterium the requirement that the overall orientation of a water molecule most be the one intuitively reasonable; for example, the O (of H_2O) and the O(of PO_4^-) internuclear distance must be larger than the H (of H_2O) to O (of PO_4^-) distance.

In Tables 52 and 53 we present the water molecules strongly hydrogen bonded to the sugar units and to the bases, respectively. The absence of solvating water molecules at some of the sugar units (or bases) is a consequence of the above restrictive definitions. By relaxing the hydrogen bond length criteria, more molecules would be assigned to a given group of B-DNA.

With these restrictive definitions and considering only strongly hydrogen

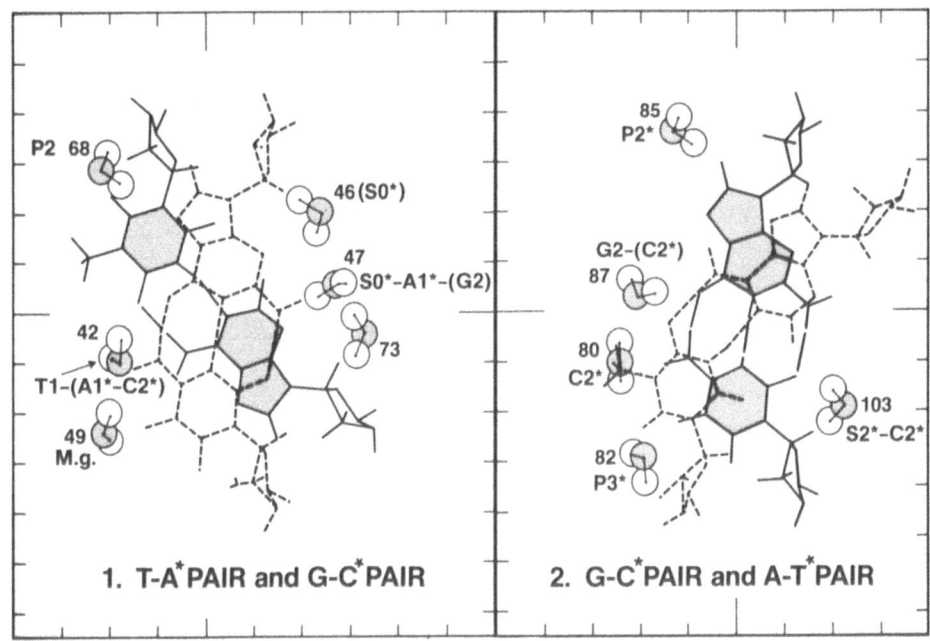

Figure 62. Water molecules in the vicinity of T1-A1* and G2-C2* (insert 1) and G2-C2* and A3-T* (insert 2).

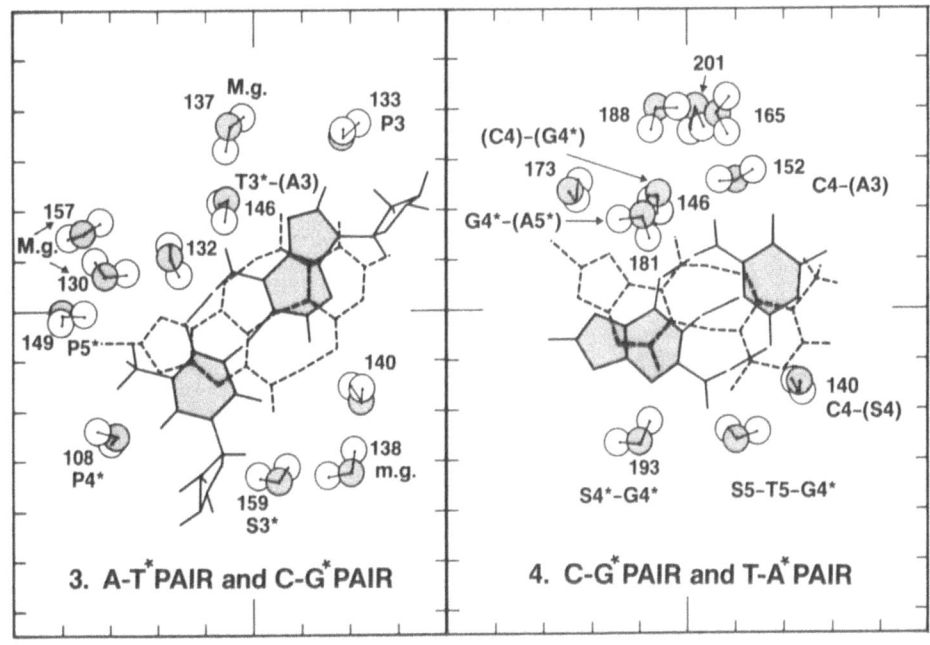

Figure 63. Water molecules in the vicinity of A3-T3* and C4-G4* (insert 3) and C4-G4* and T5-A5* (insert 4).

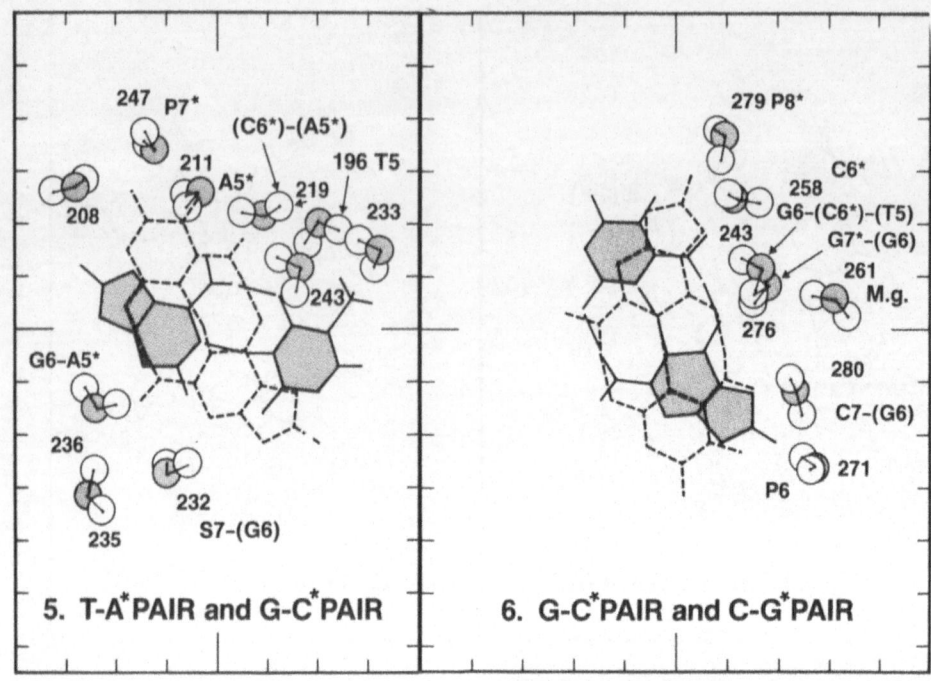

Figure 64. Water molecules in the vicinity of T5-A5* and G6-C6* (insert 5) and G6-C6* and C7-G7* (insert 6).

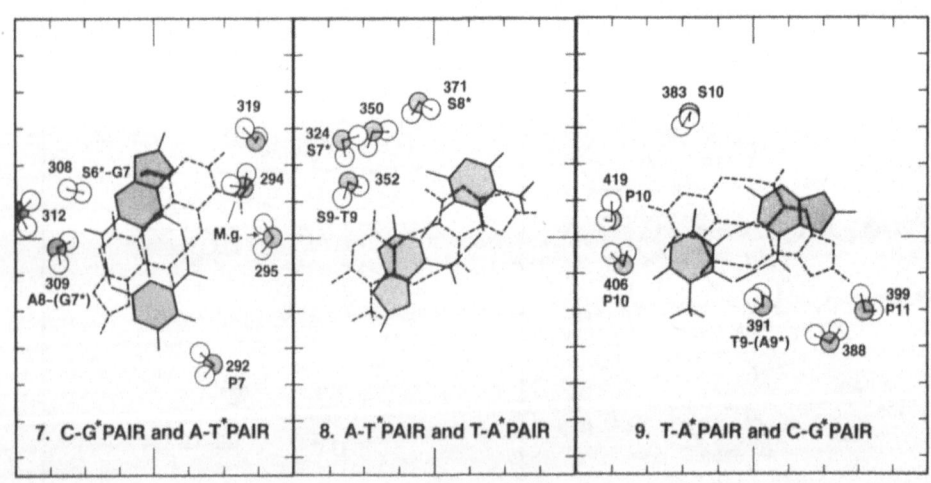

Figure 65. Water molecules in the vicinity of C7-G7* and A8-T8* (insert 7) and A8-T8* and T9-A9* (insert 8) and T9-A9* and C10-G10* (insert 9).

bonded water molecules, we can summarize as follows: there are 5.1 water molecules solvating each PO_4^- group, 0.7 water molecules solvating each sugar group, 0.5 water molecules solvating both the sugars and bases (namely, hydrogen bonded bridges between a sugar and a base) and

Table 52. Water Molecules Solvating The Sugar Units.

S#	W#	R(H1)	R(H2)	E(W-DNA)	E(W-W)	Notes
S2	46	1.9	3.3	-98.1±4.6	-9.2±4.7	S2
S5	179	1.9	3.1	-106.2±4.3	-10.6±2.6	S5-T5-G4*
S7	232	2.2	3.7	-90.8±2.9	-11.5±2.6	S7-(G6)
S8	309	1.7	3.0	-89.5±5.5	-6.2±4.2	S8-A8-(G7*)
S9	352	1.8	3.0	-84.3±4.1	-15.1±2.5	S9-T9
S10	383	1.9	2.9	-78.7±4.2	-16.9±3.2	S10
S0*	47	2.0	3.2	-48.4±6.9	-23.3±3.5	S0*-A1*-(G2)
S2*	103	1.8	2.9	-89.4±3.9	-16.0±4.7	S2*-C2*
S3*	159	1.7	3.0	-84.8±3.1	-19.3±4.8	S3*
S4*	193	2.0	3.0	-81.9±5.7	-14.4±4.4	S4*-G4*
S5*	236	1.9	3.1	-100.5±2.9	-9.4±3.9	S5*-G6-A5*
S6*	308	2.0	3.4	-92.0±3.3	-7.2±2.8	S6*-G7
S7*	324	1.8	2.9	-89.8±3.8	-4.8±2.8	S7*
S8*	371	2.0	3.3	-83.6±5.2	-12.3±5.1	S8*
AV		1.9	3.1	-86.9±4.3	-12.6±3.7	

Table 53. Water Molecules Solvating The Base-Pairs

W#	BASE	R(1)	ATOM(1)	R(2)	ATOM(2)	E(W-DNA)	E(W-W)	Notes
42	T1	1.9	O4			-70.5±5.8	-12.0±3.5	T1-(A1*)-(C2*)
	A1*	2.6	HN61					
	C2*	3.0	HN42					
47	A1*	2.2	N3			-48.4±6.9	-23.3±3.5	S0*-A1*-(G2)
	G2	2.8	HN21					
80	C2*	1.9	HN42			-65.5±2.9	-15.8±4.7	C2*
87	G2	1.9	O6			-75.5±5.1	-7.5±6.1	G2-(C2*)
	C2*	2.8	HN41	3.0	HN42			
103	C2*	2.0	O2	3.0	O2	-89.4±3.9	-16.0±4.7	S2*-C2*
132	T3*	2.0	O4			-75.6±4.3	-6.3±5.2	T3*-A(A3)
	A3	2.6	HN61	2.6	HN62			
140	C4	1.9	O2	3.0	O2	-81.9±3.4	-13.8±4.3	C4-(S4)
146	C4	2.6	HN42			-61.0±3.3	-17.8±4.7	(C4)-(G4*)
	G4*	2.9	O6					
152	C4	1.9	HN42			-55.0±2.8	-26.5±3.9	C4-(A3)
	A3	2.7	N7					
179	T5	2.0	O2	2.9	O2	-106.2±4.3	-10.6±2.6	S5-T5-G4*
	G4*	2.3	HN21					
181	G4*	1.9	O6			-79.9±4.7	-18.0±3.3	G4*-(A5*)
	A5*	2.9	N7	2.7	HN61			
193	G4*	2.3	N3			-81.9±5.7	-14.4±4.4	S4*-G4*
196	T5	2.3	O4			-55.2±4.1	-12.3±4.4	T5
211	A5*	2.3	HN61			-60.6±3.3	-27.3±5.1	A5*
219	A5*	3.0	HN61			-26.0±7.8	-33.9±5.0	(C6*)-(A5*)
	C6*	2.8	HN42					
232	G6	2.5	N3	2.6	N3	-90.8±2.9	-11.5±2.6	S7-(G6)
236	G6	2.3	HN21			-100.5±2.9	-9.4±3.9	S5*-G6-A5*
	A5*	2.3	N3					
243	G6	2.0	O6	2.8	O6	-69.9±4.3	-10.7±2.5	G6-(C6*)-(T5)
	C6*	2.8	HN42					
	T5	2.8	O4	2.8	O4			
258	C6*	1.9	HN42			-26.0±3.0	-31.7±5.0	C6*
276	G7*	1.9	O6			-69.4±3.7	-16.1±2.8	G7*-(G6)
	G6	2.5	O6	2.9	O6			
280	C7	1.8	HN42			-77.2±3.7	-14.6±2.8	C7-(G6)
	G6	2.4	N7					
308	G7*	2.8	N3	2.3	N3	-92.0±3.3	-7.2±2.8	S6*-G7
309	A8	2.2	N3			-89.5±5.5	-6.2±4.2	S8-A8-(G7*)
	G7*	2.4	HN21					
352	T9	2.0	O2	2.7	O2	-84.3±4.1	-15.1±2.5	S9-T9
391	T9	1.9	O4			-83.4±3.9	0.9±5.0	T9-(A9*)
	A9*	2.8	HN61					

0.9 water molecules solvating a base. These average values refers only to the first solvation shell, in the strict sense above defined; in this way, only 140 water molecules out of a total of 447 are considered. The average water-B-DNA interaction energies (in kJ/mole) are -104.7 ± 5.7, -86.9 ± 4.3, -85.9 ± 4.4 and -63.4 ± 4.2 for the PO_4^-, sugar, sugar and base, bases, respectively; the average water-water interaction energies

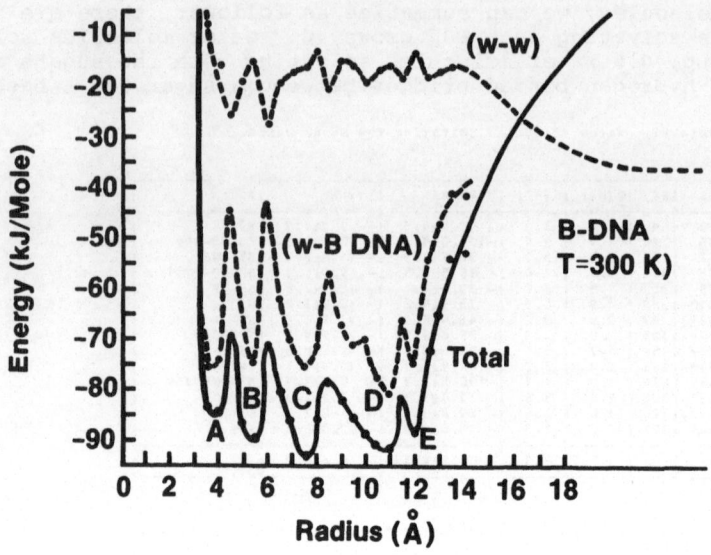

Figure 66. Interaction energies of a water molecule with B-DNA in kj/mole.

are -5.3 ± 3.8, -12.6 ± 3.7, -12.6 ± 3.5 and -16.6 ± 4.3 kJ/mole for the same groups above listed. Let us now consider the energetic of the full set of 447 water molecules (see Figure 66).

The interaction energies (water-water, water-B-DNA and total) are reported for the water molecule enclosed in the interface of two co-axial cylinders differing in the two radii by 0.5A. The first point (R = 0.5) refers to the energy of the water molecules enclosed in a cylinder of radius R = 3.5A (it contains no water molecules). Extrapolation from R = 14.5A to very large R qualitatively is easy, since the water-DNA interaction will go to zero and the water-water interaction will go to the simulated value for bulk-water (35.6 kJ/mole, see Reference 90). Quantitatively however, the extrapolation is somewhat more difficult and we would like to limit ourselves to present the simulated data, namely for a water molecule from R = 0 to R = 14.5A the average total energy is -79.5±0.1 kJ/mole, and the average water-water interaction energy is -16.3∓0.05 kJ/mole; the total energy simulated value is an upper limit to the water-DNA interaction energy. Extension of our simulation to about 1500 to 2000 water molecules would fully settle this point, but being the corresponding simulation presently somewhat too expensive, we have deferred it to a later date. The total energies above given can be somewhat misleading since obtained as an average over many different positions, with large energy variations. The total interaction energy of Figure 66 clearly shows 5 minima corresponding to interactions with the bases (A), bases and sugars (B), sugars (C), and phosphates (D and E). The energy curve is computed up to 14.5A and "freely" extrapolated thereafter. The water-water (W-W) and the water-DNA (W-BDNA) components of the interaction energy support the above interpretation for the energy minima pattern; however, the partial interaction energies are more structured in the region from 8A to 14.5A; the

3 to 4 minima, seems to provide a differentiation between the four oxygen atoms of PO_4^-. This analysis, however, is complicated by the fact that much water is located in the major and in the minor grooves. Such water has little resemblance to bulk water since it is highly structured, and therefore its energy is not a "constant additive" contribution to the energy curve.

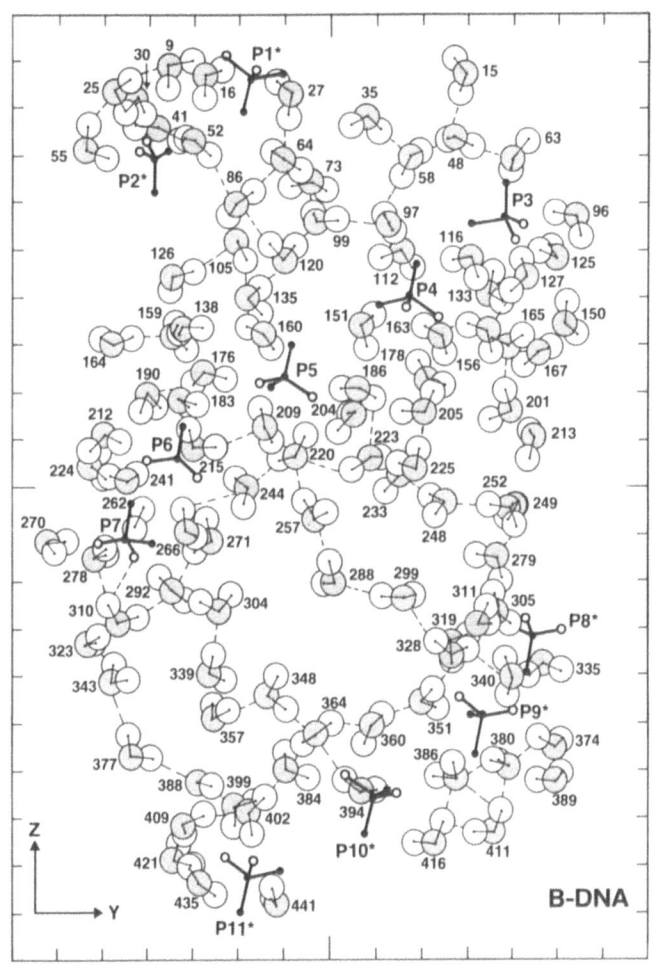

Figure 67. Network of water molecules in B-DNA (see text).

The complexity of the structure of water in the two grooves is evident in Figure 67, where we consider the water molecules enclosed in the volume between two co-axial cylinders of radii 8A and 12A, respectively; the figure reports the water of only one half of the cylinder volume on a y, z projection. To us the striking features of this figure is the clear evidence of hydrogen bonded water filaments from a phosphate group of h to a phosphate group of h*, spanning the major groove (the filaments have nearly periodic roto-translational symmetry) and of hydrogen bonded water filaments connecting two successive phosphate groups, from P(i) and P(i+1) in the h (and/or in h*) helix. We have previously commented on these <u>filaments</u> (189, 190, 192, 194) since either implicit in the iso-energy maps or explicit in the A-DNA and B-DNA single helix Monte Carlo simulations; however the limited number of water molecules considered in our previous simulations was limiting the validity of our suggestion. We note that this feature - the filaments existence - has been previously encountered in ion-pairs in solution (70). We feel that this feature is basic in any water solution containing ions, and will have profound consequences to the understanding of dynamical and temperature dependent properties of solution containing ions.

The hydrogen bonds reported in the figure are reported only if the oxygen-oxygen distance (between two waters) is equal or smaller than 3.5A, and if the oxygen hydrogen distance is smaller than the corresponding oxygen-oxygen distance. Typical water filaments (see Figure 67) are formed by the waters 310, 323, 343, 377, 388 and 399 linking P7 of h to P11* of h*; or 271, 266, 292, 304, 339, 357, 348, 364 linking P6 of h to P10* of h*; or 220, 257, 288, 299, 328 and 340 linking P5 of h to P9* of h*. These are <u>transgroove</u> filaments. Other structured filaments are present and connect a phosphate to a successive phosphate in the same helix. Notice for example, the water molecules for the <u>inter-phosphate</u> filaments 402, 384, 364, and 394 from P11* to P10* and 364, 360, 351 from P10* to P9*. In the above examples waters 310 and 399 are <u>terminal waters of a transgroove filament</u>; the structure in the <u>inter-phosphate</u> filaments is different since the filament 402, 384, 369 (from P11* to P10*) <u>continues</u> with waters 360, 351 leading to P9*. Being the network complex, there is some element of arbitrariness in defining terminal waters in a filament; however the finding of two different structural organizations, namely the <u>transgroove</u> filaments and the <u>inter-phosphate</u> filaments, seems firm. It is stressed that these structures do not correspond to data obtained by analyzing <u>one</u> or <u>few</u> conformation, but are statistically "stable" and meaningful structures. It is very tempting to postulate that proton "are transferred preferentially along these filaments"; hence, these filaments are of importance in reactivity studies. These structures are "dynamical" in the sense that a given structure can evolve into a different structure, at relatively little expense for the <u>total</u> energy of the system (these findings are related to the comments at the end of Section 2.3).

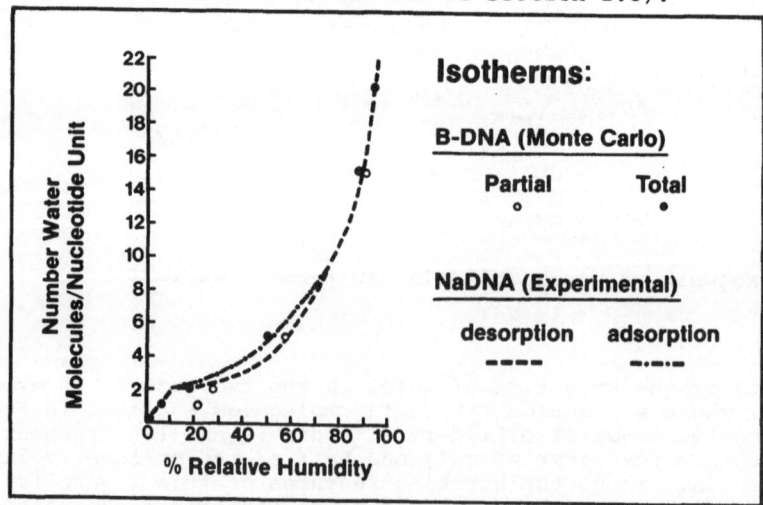

Figure 68. Monte Carlo simulated isotherm at 300°K for B-DNA and Na-DNA experimental isotherms at 300°K.

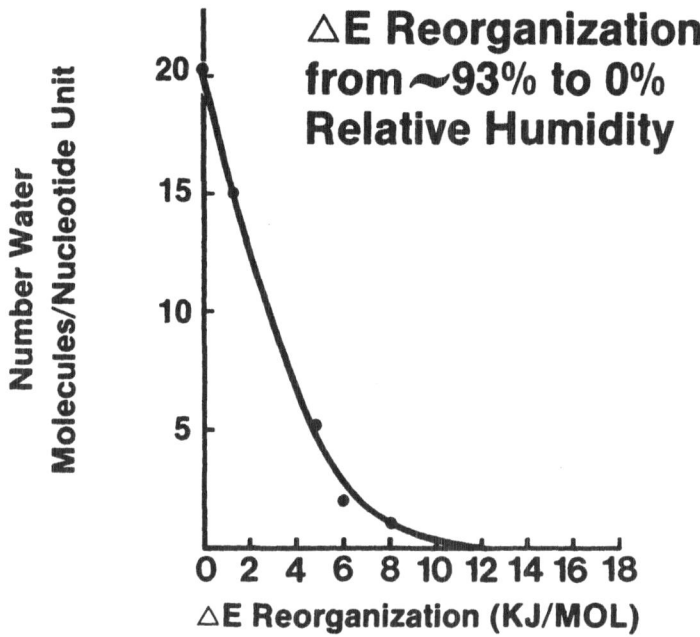

Figure 69. Reorganization energy of water molecules relative to a sample with 92% water humidity (in B-DNA) at 300°K (see text).

Having briefly analyzed these micro-aspects of the DNA solvation, let us now turn to the other extreme and analyze some macro-aspect, in particular, the absorption and desorption isotherms. These have been studied experimentally in depth (195 to 201). We refer, in addition, to a short but recent review on the paper by J. Texter on this point (202). The absorption-desorption hysteresis cannot be obtained directly from Monte Carlo simulations, if the solute is not allowed to structurally adjust itself to concentration variations of the solvent. However, the hysteresis is a "fine details" of the isotherm; the isotherm main characteristic is the well known sigmoidal shape. We note that our simulation (447 water molecules) corresponds to 92% of relative humidity, or 20.3 water molecules per nucleotide unit. If we start with 447 water molecules and subtract progressively an increasing number of water molecules (desorption simulation) without allowing the water molecules to rearrange, the simulated isotherm is not sigmoidal but nearly linear. If however we subtract water molecules and we reperform a Monte Carlo simulation then the water molecules can rearrange themselves: in this case the isotherm has a nicely sigmoidal shape (see Figure 68). By construction (see above) the simulated desorption and absorption isotherms are equal. In Figure 68 we have reported also the desorption and absorption isotherms for Na-DNA. Clearly, the two experiments (our simulation and the Na-DNA isotherms) should not be compared, attempting to find a one to one correspondence. In Figure 69 we report the water reorganization energy for the water molecules solvating B-DNA, namely the energy difference between a Monte Carlo simulation with N water molecules (with N<447) and the energy of the N water molecules taken from the sample of the 447 water molecules. From the figure it is evident that reorganization energy increases by decreasing the number of molecules in the solvent, as clearly expected. In our new simulation we have selected the following values for N: 333, 114, 44, and 22. Lowering the percentual humidity brings about first elimination of the water weakly bound (from large R values to R = 13A), then elimination of the <u>trans-groove</u> filaments, then elimination of water from the vicinity of the bases and from the sugar and finally from the <u>inter-phosphate</u> filaments.

Removal and/or rearrangement of the trans-groove filaments brings about a rearrangement of the B-DNA structure, that in turn induces a second water rearrangement proportional to the width of the hystersis loop. Therefore, simulation experiments of the type here reported <u>coupled</u> with laboratory isotherms experiments will allows to differentiate between water rearrangements and DNA induced rearrangement. Unfortunately, this comparison cannot be made with B-DNA without counter-ions.

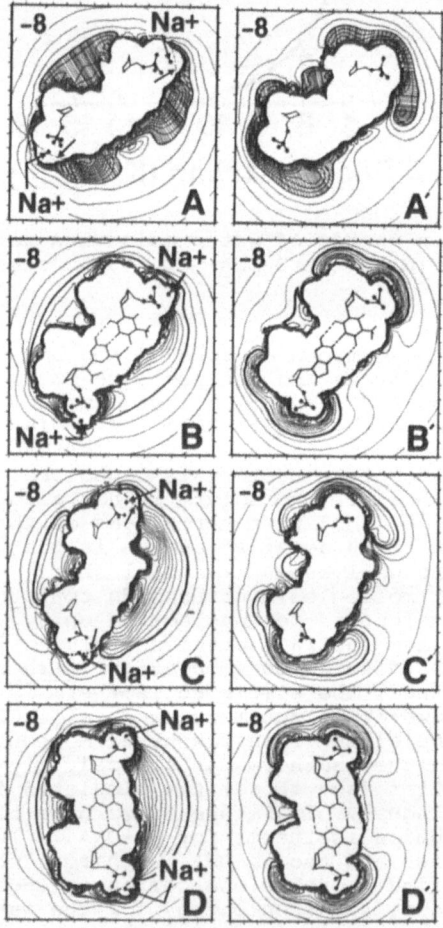

<u>Figure 70</u>. Iso-energy maps for the interaction energy of one water molecule with B-DNA (right) and Na^+-B-DNA (left).

Let us now shortly comment on some aspect of the solvation of Na^+-DNA limiting ourself to qualitative reasoning based on iso-energy maps presented in Figure 70. The 22 phosphate groups of our B-DNA fragment have been neutralized by placing a Na^+ ion at the energy minimum position of Na^+ interacting with diethyl phosphate and other phosphates (184, 185 and 203). The four inserts A, B, C and D on the left of Figure 70 correspond to iso-energy maps for a water molecule interacting with the fragment of B-DNA having one Na^+ ion at each phosphate (hereafter referred to as Na^+-B-DNA, for short). The four planes of the inserts are mutually parallel and contain the phosphate groups above the A-T* pair (insert A), the A1-T1* pair (insert B), the phosphate group above the G2-C2* pair (insert C) and the G2-C2* pair (insert D). The iso-energy maps in the four inserts to the right (A', B', C' and D') are in corresponding planes for B-DNA. The contour corresponding to -8.0 Kcal/mole (outmost contour) is explicitly indicated; the contour to contour energy difference is 2.0 kcal/mole. Comparing inserts A and A' we notice that the area delimited by the contour at -14.0 kcal/mole and by the hard core (the most attractive region, shaded areas in Figure 70, inserts A and A') increases in the vicinity of the sugars and bases in A relative to A'. The same feature is evident by comparing B with B', C with C' and D with D'. In the maps B, B', C, C', D, and D' the contour at -14.0 kcal/mole is distinguished by heavier lines (relative to the remaining contours). As a consequence of the energetic variations in the most attractive regions we expect that the water molecules in the minor and major grooves will be more attracted in Na^+-B-DNA than in B-DNA; the same comment holds for the water solvating the base-pairs. The _large_ difference in the contour maps between B-DNA and Na^+-B-DNA ensure that there are corresponding differences in the water structure, namely a) in the number of water molecules solvating the first shell and, b) in the structure of the hydrogen bonded filaments, previously discussed. We can restate this point in a different way. The type of reasoning presented to explain the hysteresis in the isotherms is here advanced to support the expectation of induced rearrangements in the DNA structure. However, the iso-energy contour differences between Na^+-B-DNA and B-DNA are _very_ large; therefore, we expect _significant_ variations in the DNA structure, in agreement with the experimentally well known transitions from one conformation of DNA to another, induced by variations in the concentration and/or type of counter-ions.

5.13 SOLVATION OF Na^+-B-DNA AT 300°K

Here we report on a preliminary Monte Carlo study on the solvation of the previously discussed fragment of B-DNA neutralized with one Na^+ ion at each PO_4^- group. The positions for the Na^+ ions are those discussed in the previous section (5.12). The number of water molecules simulated in the Monte Carlo computation is 447; the simulated temperature is 300°K (204).

It is noted that by assuming a fixed Na^+ position we introduce an element of arbitrariness. Indeed in computations on the zwitterionic form of glycine, enclosed in a cluster of 200 water molecules and in presence of one Na^+ion, we have found (205) that the Na^+ ion has a rather high probability to be at any position within a sphere of about 0.5 A radius. Therefore, a more realistic approach would let the Na^+ ions free to move (the most probable positions should be determined by the Monte Carlo method itself). On the other hand, if we consider the Na^+-B-DNA macromolecule as our solute, rather than B-DNA (in a solution of water and Na^+ ions), then vibrational freedom should be given not only to the Na^+ ions, but also to the whole Na^+-B-DNA macromolecule (work in this direction is in progress).

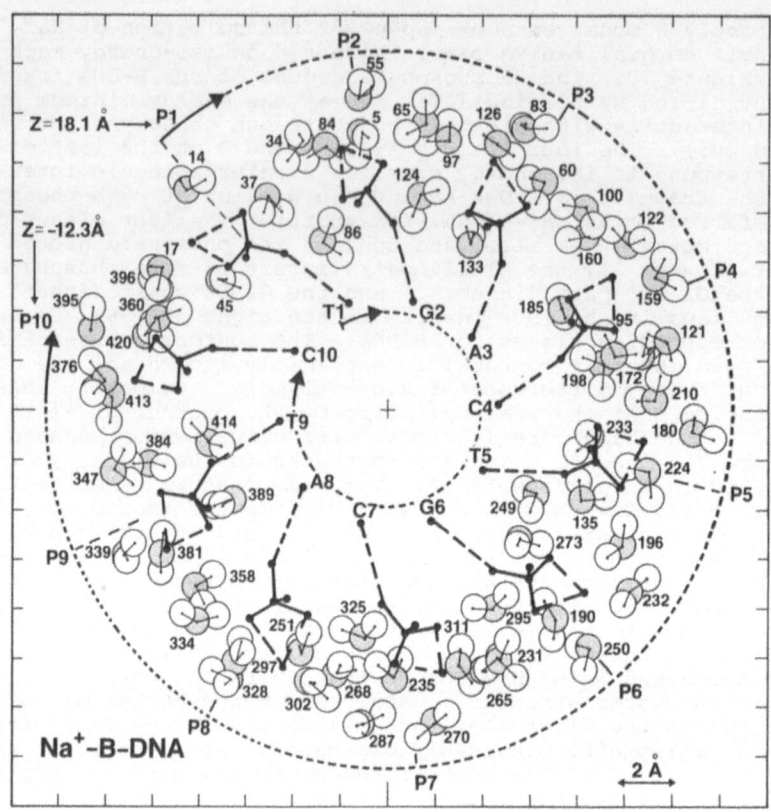

Figure 71. Water molecules solvating the h helix in Na$^+$-B-DNA double helix.

With this in mind, let us consider the results of a simulation on Na$^+$-B-DNA, carried out in a way very similar to the previously reported simulation on B-DNA (see Section 5.12).

In Figure 71 re report the water molecules solvating the Na$^+$-PO$_4^-$ groups on the h helix of the double-helix. We have reported those water molecules that are <u>strongly</u> bound according to the stringent definitions of "strong" hydrogen bounding previously given (Section 5.12). Would we consider also "intermediate strength" hydrogen bonds, we would have to include many more water molecules. Comparing the water organization in Figure 71 with the equivalent situation for B-DNA, we immediately note the larger number of water molecules in the Na$^+$-B-DNA relative to B-DNA; in addition, the water molecules penetrate more deeply (shift towards smaller R values of the cylinder containing the water molecules). In Tables 54 to 57, we report the water molecules strongly hydrogen bonded to PO$_4^-$, either in the h helix (Table 54) or in the h* helix (Table 55) and those strongly hydrogen bonded to Na$^+$ either in the h helix (Table 57) or in the h* helix (Table 56).

In the Figures 72 to 75 we report the water molecules strongly solvating the base pairs. As previously done in the B-DNA analyses, in each figure we present two base-pairs; the one drawn with full lines lies above (higher z) to the one drawn with dashed lines. The water molecules strongly hydrogen bonded to the sugar units in the h and h* helices, and to the bases are reported in Tables 58 and 59, respectively. These

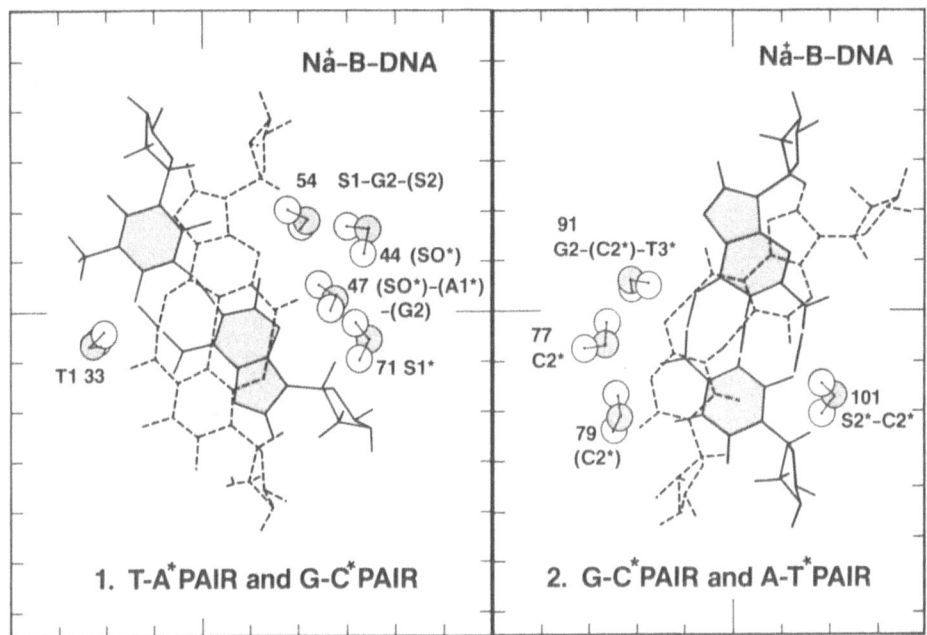

Figure 72. Water molecules bonded to T1-A1*, G2-C2* (insert 1), G2-C3, A3-T3* (insert 2).

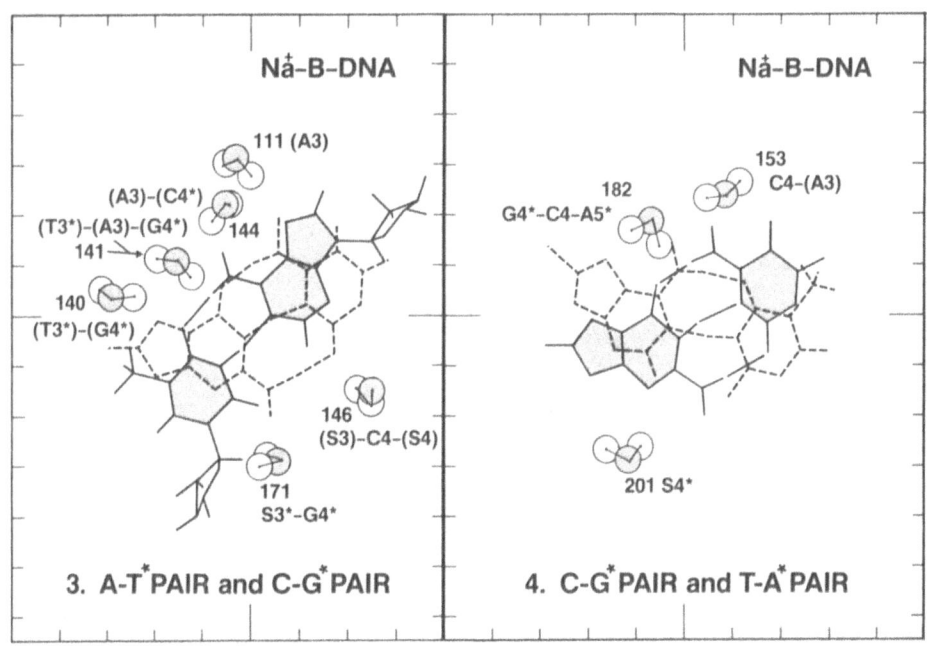

Figure 73. Water molecules bonded to G2-C2, A3-T3* (insert 3), G4-C4*, T5-A5* (insert 4).

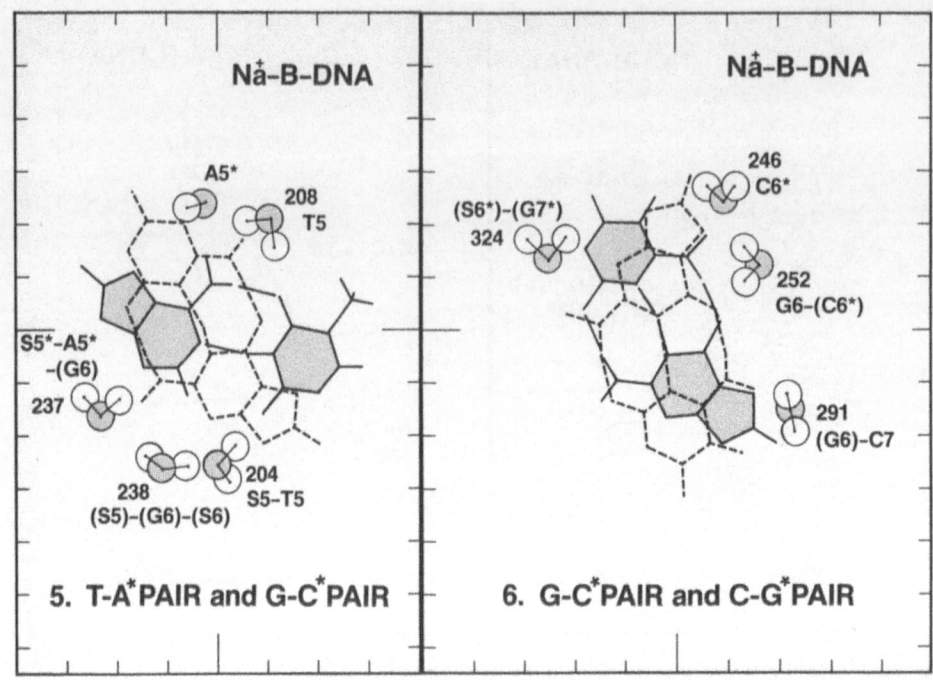

Figure 74. Water molecules bonded to G4-C4*, T5-A5* (insert 5), G6-C6*, C7-G7* (insert 6).

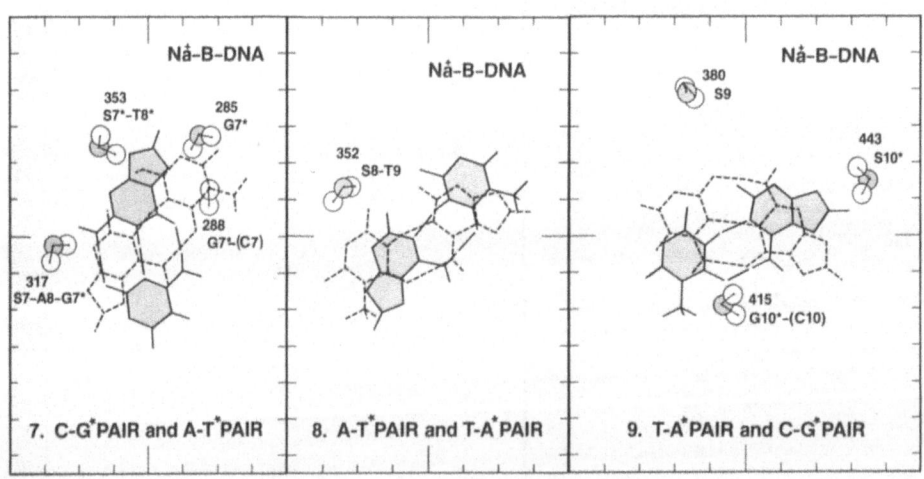

Figure 75. Water molecules bonded to G6-C6*, C7-G7* (insert 7), A8-T8*, T9-A9* (insert 8), T9-A9*, C10-G10* (insert 9).

Table 54. Water Molecules Solvating PO_4^- in the h Helix of Na^+-B-DNA*.

P#	W#	Atom(1)	R(H1)	Atom(2)	R(H2)	E(W-Na-B-DNA)	E(W-W)
P1	14	O1P	2.1	O1P	3.3	−140.9±5.0	1.0±4.5
	37	O1P	2.0	O1P	3.5	−131.3±5.7	−8.6±3.2
	45	O2P	1.6	O2P	3.0	−119.5±5.7	−13.8±3.2
P2	65	O1P	3.0	O1P	1.6	−117.8±6.8	−12.8±4.1
	86	O2P	3.1	O2P	1.8	−123.6±7.6	−22.4±5.1
	5	O3'	3.2	O3'	1.8	−106.3±5.0	−6.6±3.8
P3	100	O1P	1.6	O1P	3.0	−132.5±5.6	−18.3±4.8
	124	O2P	2.7	O2P	1.7	−128.8±8.8	−17.2±4.3
	133	O2P	1.9	O2P	3.1	−105.4±3.8	−28.7±4.5
	60	O3'	2.5	O1P	2.1	−86.0±7.1	−19.8±3.2
P4	121	O1P	2.9	O1P	1.7	−111.4±6.5	−24.8±3.5
	172	O1P	3.1	O1P	1.7	−127.0±11.6	−23.9±3.4
	185	O2P	3.1	O2P	1.8	−114.6±5.0	−28.6±3.4
	95	O3'	3.2	O3'	2.1	−79.1±8.5	−30.3±2.4
P5	196	O1P	1.7	O1P	3.0	−145.2±3.5	−12.4±4.5
	198	O2P	3.0	O2P	1.7	142.2±3.7	−15.3±4.1
	233	O2P	1.6	O2P	3.0	−136.9±4.2	−12.4±4.5
	135	O3'	1.9	O3'	2.9	−90.5±3.6	−30.2±3.1
P6	231	O1P	2.8	O1P	1.6	−127.6±4.3	−23.8±2.6
	249	O2P	2.8	O2P	1.7	−140.8±6.8	−12.9±7.4
	273	O2P	2.7	O2P	1.8	−104.1±4.5	−33.9±3.1
	190	O3'	3.1	O3'	1.7	−105.1±5.8	−12.1±3.0
P7	268	O1P	2.8	O1P	1.5	−132.4±4.2	−6.6±3.2
	295	O2P	1.7	O2P	3.2	−142.1±6.3	−10.9±3.5
	235	O3'	3.1	O3'	1.8	−116.2±5.0	−14.3±3.6
	287	O1P	2.1	O1P	3.0	−103.6±8.4	−12.0±3.6
P8	297	O1P	1.9	O1P	3.3	−141.6±4.6	−6.4±4.1
	334	O1P	3.0	O1P	1.6	−139.5±4.0	−11.2±3.5
	325	O2P	3.0	O2P	1.8	−144.4±5.8	−4.9±3.6
	351	O2P	1.9	O2P	3.3	−168.4±5.5	−0.5±4.5
P9	347	O1P	1.5	O1P	2.7	−123.8±4.9	−17.5±6.3
	384	O1P	1.9	O1P	2.8	−119.9±5.5	−15.3±5.0
	358	O2P	3.4	O2P	2.1	−153.2±5.9	−1.8±4.5
	389	O2P	3.0	O2P	1.7	−103.1±4.1	−22.8±5.1
P10	396	O1P	1.8	O1P	2.7	−109.4±4.8	−12.1±2.9
	420	O1P	1.6	O1P	3.1	−120.9±5.9	−18.1±3.9
	414	O2P	1.9	O2P	2.9	−113.6±6.1	−13.1±5.0
	360	O3'	2.8	O3'	1.7	−111.5±3.8	−28.8±3.3
P11	442	O1P	1.5	O1P	3.0	−117.9±2.9	−9.4±5.1
	439	O2P	1.5	O2P	2.9	−138.9±2.6	−15.3±5.1
	399	O3'	3.2	O3'	1.8	−101.7±3.0	−19.6±3.7

*The H1 to "atom 1" and H2 to "atom 2" distances are in Å unit; energies in kJ/mole.

tables are to be compared with the corresponding one for B-DNA. We can summarize by stating that 6.8 water molecules are strongly hydrogen bonded to each Na^+-PO_4^- group (3.8 solvate Na^+, 3.0 solvate PO_4^-), 0.6 are strongly hydrogen bonded to each sugar group and 1.2 water molecules are strongly hydrogen bonded to each base. The exact meaning of these average values should be taken with reference to Tables 54 to 59. Once more we stress that a given statistical position and orientation for a water molecule is due to the <u>entire</u> field of the macro-molecule and of the solvent, therefore the existence of a water molecule at a given site cannot be attributed only to the field created by the DNA atoms at that site. Note that the O5' oxygen is PO_4^- is not solvated because of the CH_2 field (but the O3' is solvated). This finding points out some reason of concern of studies limited to symmetrical model compounds, such as diethyl phosphate, if assumed as a prototype of PO_4^- in nucleic acids.

In Figure 76 we report the total, the water-water and the water Na^+-B-DNA interaction energies vs. the radius of the cylinder (containing the water molecules). Comparing this figure with the equivalent one for B-DNA (see Section 5.12), it is apparent that, a) there is an overall energy lowering, b) a shift towards smaller values of R and, c) the same overall structure of minima and maxima.

The isotherm for Na^+-B-DNA has been computed in a way equivalent to the simulation for B-DNA. The obtained sigmoidal shape closely resembles the one simulated for B-DNA; there are however displacements, especially in the low humidity regions, as expected due to the different number of strongly hydrogen bonded water molecules to PO_4^- (in B-DNA) and in Na^+-PO_4^- (in Na^+-B-DNA).

Table 55. Water Molecules Solvating PO_4 in the h* helix of Na^+-B-DNA*.

P*#	W#	Atom(1)	R(H1)	Atom(2)	R(H2)	E(W-Na-B-DNA)	E(W-W)
P1*	30	O1P	2.8	O1P	1.8	-110.4±4.7	-19.4±3.7
	1	O2P	1.8	O2P	3.0	-118.5±6.4	-7.7±3.2
	63	O3'	1.8	O3'	3.1	-109.1±3.2	-18.1±3.7
P2*	35	O1P	2.9	O1P	1.7	-126.4±5.1	-13.8±5.8
	81	O1P	2.8	O1P	1.6	-125.8±5.1	-15.5±4.1
	26	O2P	3.0	O2P	1.7	-122.7±4.6	-21.5±3.7
	104	O3'	1.8	O3'	3.1	-111.0±3.6	-22.3±4.9
P3*	105	O1P	3.0	O1P	1.6	-134.5±5.0	-18.6±4.3
	85	O2P	2.7	O1P	1.7	-113.1±4.7	-29.3±4.0
	147	O3'	1.7	O3'	2.9	-104.8±4.5	-30.9±3.7
P4*	112	O1P	2.6	O1P	1.9	-112.1±5.6	-13.9±5.4
	156	O1P	1.7	O1P	3.0	-128.4±5.7	-17.2±2.7
	109	O2P	2.8	O2P	1.5	-109.6±6.3	-15.9±5.0
	131	O2P	2.0	O2P	3.1	-141.3±3.7	0.0±3.6
P5*	158	O1P	3.4	O1P	1.9	-133.7±4.8	-11.8±3.4
	207	O1P	3.0	O1P	1.7	-121.2±4.7	-22.9±2.2
	152	O2P	2.1	O2P	3.4	-130.6±5.0	-6.3±3.4
	175	O2P	2.4	O2P	1.8	-110.6±4.3	-21.9±3.8
	225	O3'	3.5	O3'	2.1	-131.8±8.1	-9.5±6.1
P6*	209	O1P	1.9	O1P	2.9	-113.4±6.8	-25.2±3.5
	243	O1P	1.7	O1P	2.9	-128.1±4.3	-7.0±4.0
	214	O2P	1.6	O2P	2.9	-131.3±4.5	-18.0±3.8
	279	O3'	3.1	O3'	2.1	-96.3±2.5	-22.4±3.3
P7*	278	O1P	2.9	O1P	1.6	-115.9±6.0	-21.0±3.4
	254	O2P	1.5	O2P	2.6	-110.2±5.7	-23.1±3.8
	316	O3'	3.1	O3'	2.1	-107.5±4.9	-12.6±3.9
P8*	301	O1P	1.7	O1P	2.9	-145.4±3.3	-11.5±1.4
	245	O1P	2.7	O1P	1.5	-84.8±2.0	-23.1±2.7
	289	O2P	1.8	O2P	2.7	-134.0±5.3	-20.9±3.1
	312	O2P	1.8	O2P	2.9	-136.5±5.1	-19.1±5.1
	361	O3'	3.1	O3'	1.9	-115.6±4.1	-19.2±6.1
P9*	342	O1P	3.0	O1P	1.7	-131.1±6.9	-17.0±4.2
	379	O1P	2.8	O1P	1.6	-108.0±4.2	-20.9±3.8
	331	O2P	2.9	O2P	1.6	-121.8±4.1	-22.3±3.1
	355	O2P	2.7	O2P	1.7	-120.9±5.3	-18.3±5.1
P10*	394	O1P	2.7	O1P	2.0	-105.7±6.3	-12.4±3.7
	419	O1P	1.9	O1P	3.2	-103.0±5.1	-16.9±3.1
	362	O2P	1.7	O2P	3.1	-120.3±5.7	-13.3±3.8
	393	O2P	1.6	O2P	3.0	-143.7±4.8	-5.9±3.9
P11*	409	O1P	2.6	O1P	1.7	-106.8±14.9	-16.2±3.8
	446	O1P	1.5	O1P	2.9	-113.0±3.3	-9.0±4.5
	403	O2P	2.8	O2P	1.6	-123.9±7.6	-18.6±5.1
	423	O2P	1.9	O2P	2.5	-125.3±7.8	-5.5±3.5

*Oxygen (phosphate) to hydrogen (H_2O) maximals considered distance is 2.1 Å (see Table 54 for additional comments).

As in B-DNA the water network is a complex one with both <u>trans-groove</u> filaments and <u>interphosphate</u> filaments. From the previous tables we have learned that the following water molecules 15, 1, 35, 8, 30, 63 and 135 strongly solvate the Na^+ (of P2*), P1*, P2*, Na^+ (of P1*), P1*, P1* and P5, respectively. From an analyses of the Monte Carlo probability density we have determined many patterns of water filaments. For example, one pattern is composed by the water molecule 9 linked to 15, 15 to 1, 1 to 35, 35 to 20 and to 6, 20 to 6, 6 to 8 and to 30, 8 to 30, 30 to 63 and 48, 63 to 90, 48 to 82, 82 to 110, 90 to 110, 110 to 135, 99 and 148. In addition, 35 and 6 are linked to 50, 50 is to 87 and 87 is to 103 and 108. Clearly, this is only a <u>small portion</u> of the network, but it nicely exemplifies the <u>interphosphate</u> filaments (35, 20, 6, 8, 30 or 63) as well <u>trans-groove</u> filaments (either 8, or 30 or 63 linked to 135). The partial pattern is reported on the following page.

The water Na^+-B-DNA interaction energy, per water molecule, averaged over the 2×10^6 Monte Carlo configurations is -89.3±0.05 kJ/mole; the water-water interaction is -22.0±0.05 kJ/mole. These data are averages relative to the 447 water molecule sample; a larger sample will likely yield somewhat lower energy.

A comparison with B-DNA, Li^+-B-DNA and Mg^{++}-B-DNA is in progress; when such data will be available a comprehensive picture on the solvation aspects of nucleic acids will emerge at the micro-level (quantum mechanical aspects), at the statistical level (monte Carlo simulations) and at the thermodynamical level (for example, isotherms).

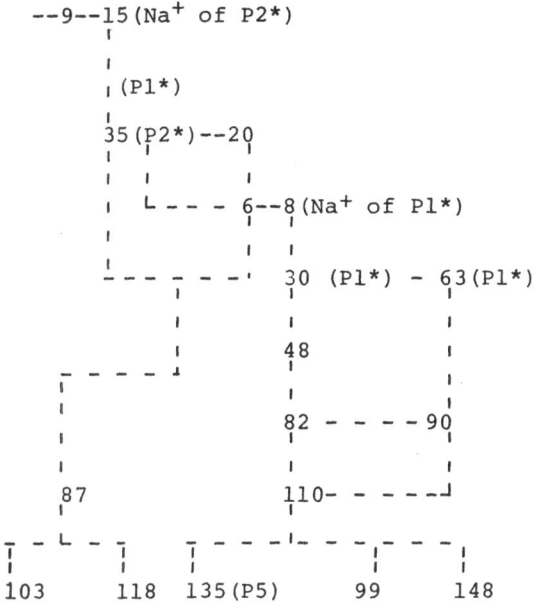

Table 56. Water Molecules Solvating the Na+ Counter Ions In the h Helix of Na+-B-DNA.

NA #	W#	R(O -NA)	Bridges	E(W-Na-B-DNA)	E(W-W)
Na(P1)	14	2.1	P1	-140.9±5.0	1.0±4.5
	17	2.0		-121.8±5.9	-13.8±3.3
Na(P2)	34	2.1		-131.6±2.6	0.6±2.6
	55	2.3		-114.3±7.7	-8.7±4.5
	84	2.1		-137.1±3.7	-13.3±3.6
Na(P3)	83	2.3		-129.4±3.5	-18.0±5.5
	97	2.1		-125.3±5.0	-13.8±3.6
	126	2.2		-128.5±3.7	-15.9±4.8
Na(P4)	122	2.2		-135.9±2.9	-9.4±2.4
	159	2.2		-105.4±3.7	-14.8±3.5
	160	2.2		-139.3±3.7	-4.7±4.2
Na(P5)	180	2.3		-110.2±6.7	-14.8±4.9
	210	2.1		-139.3±3.7	-5.6±2.8
	224	2.4		-128.2±3.6	-18.6±3.3
Na(P6)	232	2.2		-130.7±3.5	-21.5±4.0
	250	2.4		-131.2±3.6	-11.0±3.2
Na(P7)	265	2.2		-143.2±4.7	-10.5±4.2
	270	2.3		-128.8±5.3	-11.3±2.3
	311	2.4		-132.2±5.7	-16.7±3.8
Na(P8)	302	2.2		-144.9±5.6	-14.8±3.6
	328	2.2		-126.2±6.4	-6.6±3.6
	351	2.5	P8	-168.4±5.5	-0.5±4.5
Na(P9)	339	2.4		-134.4±2.5	-10.4±3.7
	358	2.1	P9	-153.2±5.9	-1.8±4.5
	381	2.4		-123.9±6.0	-13.4±4.1
Na(P10)	376	2.2		-109.5±7.6	-14.9±4.7
	395	2.7		-111.3±7.7	-16.3±2.7
	413	2.0		-123.9±3.5	-19.5±3.8
	420	2.9	P10	-120.9±5.9	-18.1±3.9
Na(P11)	410	2.4		-129.0±2.6	-16.1±3.1
	430	2.3		-101.1±9.8	-5.5±2.9
	444	2.3		-117.2±3.2	3.9±4.5

Oxygen (H_2O) to Na maximal distance considered is 3.0 Å; energies in kJ/mole.

Table 57. Water Molecules Solvating the Na+ Counter Ions in the h* helix of Na+-B-DNA

Na(P*#)	W#	R(O -NA)	Bridges	E(W-Na-B-DNA)	E(W-W)
Na(P1*)	6	2.3		-112.1±4.9	-9.6±5.1
	8	2.3		-123.1±4.6	-3.7±3.9
Na(P2*)	15	2.4		-109.9±7.4	-20.6±3.0
	43	2.4		-110.5±11.9	-8.8±5.3
	80	2.2		-134.5±2.5	-8.4±2.8
Na(P3*)	74	2.2		-130.1±5.2	-14.9±4.9
	93	2.3		-122.5±5.0	0.5±3.5
	128	2.5		-128.5±5.0	-14.9±4.9
Na(P4*)	123	2.2		-126.1±3.2	-11.8±3.8
	131	2.4	P4*	-151.3±3.7	0.0±3.6
	167	2.5		-104.9±8.5	-7.1±4.1
	169	2.4		-123.6±7.3	-10.8±3.6
Na(P5*)	158	2.8	P5*	-133.7±4.8	-11.8±3.4
	168	2.2		-132.7±6.3	-1.1±3.8
	184	2.3		-115.0±7.6	-20.7±4.5
	225	2.5	P5*	-131.8±8.1	-9.5±6.1
Na(P6*)	220	2.3		-130.4±5.4	-3.6±3.2
	239	2.1		-124.2±5.8	-7.7±3.3
	242	2.8		-132.6±6.8	-10.2±3.6
Na(P7*)	261	2.1		-137.8±5.7	-15.5±2.9
	296	2.2		-130.4±5.1	-15.1±3.6
Na(P8*)	298	2.2		-125.4±4.9	-16.9±3.4
	343	2.2		-126.1±6.1	-15.4±3.2
	322	3.0		-102.0±7.1	-14.4±3.2
Na(P9*)	348	2.2		-127.2±4.4	-12.0±3.4
	363	2.3		-140.0±3.6	-10.7±3.5
	382	2.5		-118.2±9.5	-13.3±3.6
Na	371	2.2		-131.4±2.6	-16.9±3.3
(P10*)	372	2.9		-76.9±6.1	-29.9±3.8
	412	2.1		-127.3±3.8	-18.3±3.9
Na	409	2.9	P.11*	-106.8±14.9	-16.2±3.8
(P11*)	411	2.2		-132.6±4.4	-11.3±4.5
	441	2.4		-108.9±8.5	-15.6±4.5
	441	2.1		-119.7±3.4	-8.3±3.1

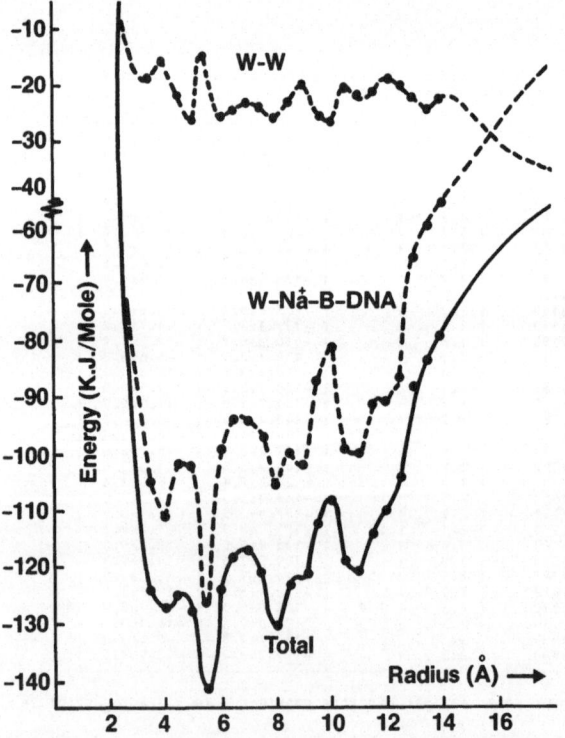

Figure 76. Interaction energies of a water molecule with Na$^+$-B-DNA.

Table 58. Water Molecules Solvating The Sugar Units in Na+-B-DNA.*

W#	Atom(1)	R	Atom(2)	R	Atom(3)	R	E(W-NaBDNA)	E(W-W)	Name
54	S1	2.0	G2	2.1	S2	2.5	-120.0±5.5	-15.3±4.6	S1-G2-(S2)
71	S1*	1.8					-115.9±2.4	-21.5±4.2	S1*
101	S2*	2.2	C2*	1.9			-132.9±3.3	-16.8±5.0	S2*-C2*
171	S3*	1.8	G4*	1.9			-129.2±3.4	-22.9±5.3	S3*-G4*
201	S4*	1.6					-134.7±3.4	-10.9±2.7	S4*
204	S5	1.9	T5	1.9			-141.8±4.2	-21.8±6.3	S5-T5
237	S5*	2.1	A5*	2.1	G6	2.3	-144.7±3.4	-7.6±4.5	S5*-A5*-G6)
317	S7	1.8	A8	2.2	G7*	2.1	-132.5±4.9	-10.9±3.8	S7-A8-G7*
352	S8	1.6	T9	2.1			-121.2±4.6	-22.6±3.8	S8-T9
353	S7*	1.7	T8*	2.1			-131.9±4.4	-14.6±4.5	S7*-T8*
380	S9	1.7					-115.0±4.0	-15.4±4.0	S9
443	S10*	1.7					-100.8±3.1	-3.1±4.5	S10*

*Energies in kJ/mole, distances in A units.

Table 59. Water Molecules Solvating Bases or Base-Pairs in Na+-B-DNA

W#	Atom(1)	R	Atom(2)	R	Atom(3)	R	E(W-NaBDNA)	E(W-W)	Name
33	T1	1.8					-101.7±2.0	-22.0±2.4	T1
77	C2*	2.1					-79.0±3.5	-30.0±3.6	C2*
91	G2	1.8	T3*	2.2	C2*	2.3	-125.1±4.2	-5.6±2.6	G2-T3*-(C2*)
146	C4	1.8	S3	2.6	S4	2.8	-130.3±3.8	-9.5±2.6	(S3)-C4-(S4)
153	C4	1.7	A3	2.4			-103.2±2.2	-22.8±3.5	C4-(A3)
182	G4*	2.1	C4	2.9	A5*	2.8	-122.7±2.5	-17.9±3.8	G4*-C4-A5*
208	T5	1.9					-119.2±1.7	-13.1±3.8	T5
222	A5*	1.9					-102.5±3.5	-24.0±50	A5*
246	C6*	2.2					-70.6±2.7	-37.4±3.5	C6*
252	G6	2.0	T	2.8	C6*	2.6	-104.3±5.3	-26.2±2.9	G6-(C6*)-(T5)
285	G7*	2.2					-81.4±4.0	-37.9±3.5	G7*
288	G7*	2.1	C7	2.9			-101.5±5.3	-21.7±4.5	G7*-(C7)
291	C7	1.8	G6	2.3			-112.5±3.4	-20.0±4.5	C7-(G6)
415	G10*	1.9	C10	2.6			-101.3±3.4	-6.9±4.5	G10*-(C10)
47	A1*	2.3	G2	2.6	S0*	2.3	-105.8±2.0	-21.6±3.8	(S0*)-(A1*)-(G2)
79	C2*	2.6					-81.2±18	-31.6±4.1	(C2*)
111	A3	2.4					-99.4±2.5	-22.5±3.4	(A3)
140	T3*	2.8	G4*	2.7			-99.6±4.5	-18.8±3.9	(T3*)-(G4*)
141	T3*	2.5	A3	2.7	G4*	2.3	-104.7±4.4	-18.5±5.0	(T3*)-(A3)-(G4*)
144	13	2.7	C4	2.9			-86.3±4.6	-30.6±5.6	(A3)-(C4)
238	G6	2.4	S5	2.7	S6	2.4	-124.1±5.1	-8.0±4.5	(S5)-(S6)-(G6)
324	G7*	2.5	S6*	2.8			-123.1±3.7	-22.5±4.5	(S6*)-(G7*)

Maximal hydrogen bond distance considered is 2.8 A; energies in kJ/mole.

5.14 CONCLUSION

About ten years ago, we have analyzed with ab-initio computations the electronic structure of one base-pair, G-C, attempting to understand the tautomeric forms induced by protons transfer from one base to the other. In Figure 77 we report the σ and the π electronic density for the G-C pair. Today we have sufficiently advanced as to consider a complex systems like B-DNA and Na^+-B-DNA in solution and at finite temperature. This progress allows us to go back and to re-analyze the

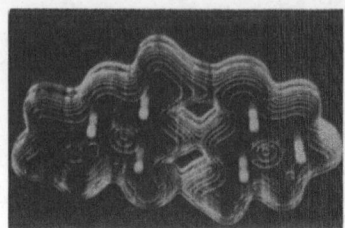 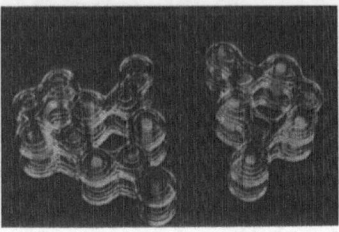

Figure 77. Electronic densities for G-C pair; σ (left) and π (right).

electronic and the vibrational structure of one or several base-pairs in the molecular field of the macromolecule and the solvent. Proton tunneling in nucleic acids, for example, represent today as much of an unanswered question as it did ten years ago. However, we can return not limited to a rather "narrow" and traditional quantum-chemical point of view, but with a broader theoretical- and computational-chemistry base, <u>demanded</u> by the many-facets nature of the chemical problem.

In a more general sense, and as conclusion of these Notes, we present in Figure 78 a matrix where each element contains either a single problem (like the determination of the structure of liquid water) or more vast problems (like electrochemistry or the determination, by simulation of the tertiary structure of proteins and/or enzymes) now more realistically open then previously by the introduction of atom-atom pair potential (as a first step), and by three or n-body potentials, obtained from ab-initio computations on model compounds. In these Notes we have only touched few applications concerning the first column of our matrix. A main task, however, for the remaining matrix elements is by now completed since the atom-atom pair potentials between most elements of the matrix are available (45) from ab-initio computations. A very extended and promising literature exist for atom-atom pair potentials obtained from experimental data; molecular dynamics simulations are becoming increasingly powerful for biological systems. It seems that simulations on complex chemical systems need neither to be limited by gross approximations nor to neglect basic parameters like field effects, temperature, time, probability distribution and entropy. In other words, the chemical problem itself determines the choice of the basic statistic to be used; therefore, for complex systems it is essential to operationally link quantum, statistical and fluid mechanics and dynamics.

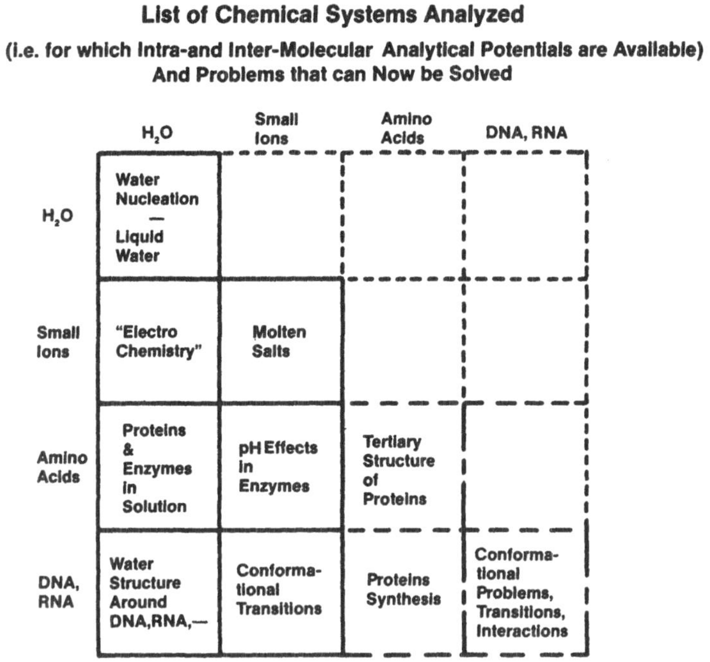

List of Chemical Systems Analyzed

(i.e. for which Intra-and Inter-Molecular Analytical Potentials are Available)
And Problems that can Now be Solved

	H_2O	Small Ions	Amino Acids	DNA, RNA
H_2O	Water Nucleation — Liquid Water			
Small Ions	"Electro Chemistry"	Molten Salts		
Amino Acids	Proteins & Enzymes in Solution	pH Effects in Enzymes	Tertiary Structure of Proteins	
DNA, RNA	Water Structure Around DNA, RNA,—	Conformational Transitions	Proteins Synthesis	Conformational Problems, Transitions, Interactions

Figure 78. Matrix of biochemical problems.

6.0 REFERENCES

1. R. S. Mulliken, J. Chem. Phys., 23, 1833 (1955); ibid., 23, 1841 (1955); ibid. 23, 2338 (1955); ibid., 23, 2343 (1955).
2. E. Clementi, Int. J. Quant. Chem. Vol. IS, 307 (1967); ibid. Vol. III S, 169 (1969).
3. The IBMOL program has been updated a number of times in the last decade. See E. Clementi and D. R. Davis, J. of Comp. Phys., 1, 223 (1966); IBM Tech. Rep. RJ #853, May (1971); RJ #883, June (1971); E. Clementi, E. Ortoleva, G. Castiglione, J. Comp. Chem. (in press).
4. E. Clementi, IBM Journal of Res. and Dev., 9 No. 1, (1965).
5. E. Clementi, L. Gianolio (unpublished data); the M.C. data are those by S. Romano and E. Clementi, Gazz. Chim. It, 108, 319 (1978).
6. E. Clementi, "Selected Topics in Molecular Physics"; Verlag Chemie Gmbh (1972).
7. E. Wigner, Phys. Rev., 46, 1002 (1934).
8. E. Clementi, J. Chem. Phys. 38, 2248 (1963); ibid. 39, 175 (1963); ibid. 42, 2783 (1965).
9. A. Veillard and E. Clementi, J. Chem. Phys. 44, 3050 (1966); ibid. 49, 2415 (1968); E. Clementi, W. Kralmer, C. Salez, J. Chem. Phys. 53, 125 (1970).
10. J. C. Slater, Phys. Rev. 81, 385 (1951).
11. W. Kohn, L. J. Sham, Phys. Rev. 140A, 1133 (1965), see also P. Hohenberg and W. Kohn, Phys. Rev. 136B, 864 (1964).
12. P. Gombas. See for example, Pseudopotentiale, Springer-Verlag, New York (1967).
13. See for example, H. Gell-Mann, K. A. Brueckner, Phys. Rev. 106, 364 (1057).
14. In addition to Reference 4, see E. Clementi in "Computers and Their Role in the Physical Sciences:, (S. Fernbach, A. Taub, Eds., Gordon and Breach 1970), pp. 503-542; E. Clementi, "Chemistry of the Cyano Group", (F. Rappoport, Ed., Wiley and Sons 1971) pp. 1-63; Proc. Natl. Acad. Sci. U.S.A., 69, 2942 (1972).
15. R. G. Gordon, Y. S. Kim, J. Chem. Phys. 56, 3122 (1972).
16. E. Clementi, B. Roos, C. Salez and A. Veillard, IBM Tech. Rep. RJ518 (1968); E. Clementi, L. Gianolio and R. Pavani, Gazz. Chim. It. 108, 181 (1978).
17. G. C. Lie and E. Clementi, J. Chem. Phys. 60, 1275 (1974); ibid. 60, 1288 (1974).
18. R. Pavani, E. Clementi, J. Chem. Phys. 67, 3403 (1977) and C. Roetti and E. Clementi, J. Chem. Phys. 60, 3342 (1974); ibid. 61, 2062 (1974). The data on the ionization potential electron affinity are to be published by Clementi, Pavani and Roetti. This study on the ionization potentials and electron affinities was reported at the III American Conference on Theoretical Chemistry, Boulder, Colorado (6/25-6/30 in 1978).
19. See for example, S. Antoci and L. Mihich, Gazz. Chim. It., 108, 383 (1978) and references therein given.
20. E. Clementi, J. of Mol. Spect., 12, 18 (1964); E. Clementi and H. Hartmann, Phys. Rev. 133, A1294 (1964).
21. See for example L. D. Landau and E. M. Lifshitz, Course of Theoretical Physics, Pergamon Press, Oxford (1960).
22. N. Metropolis, A. W. Rosenbluth, A. H. Teller and E. Teller, J. Chem. Phys., 21, 1078 (1953).
23. B. M. Alder and T. W. Wainwright, J. Chem. Phys., 31, 459 (1959).
24. See for example I. Prigogine, R. Lefever, Advances in Chemical Physics, vol. 39, R. Lefever and A. Goldbeter Eds., Wiley and Sons, New York (1978); G. Nicolis, I. Prigogine, Selforganization in Nonequilibrium System, Wiley-Interscience, New York (1977); R. Glansdorff, I. Prigogine, Thermodynamic Theory of Structure, Stability and Fluctuations, Wiley-Interscience, New York (1971).

25. K. S. Pitzer, J. Chem. Phys., 5, 469 (1937).
26. K. S. Pitzer and W. D. Gwinn, J. Chem. Phys., 10, 428 (1942).
27. H. Lifson, J. Chem. Phys., 30, 964 (1959).
28. A. M. Liquori, "Principles of Biomolecular Organization" Ciba Foundation Symposium, London, p. 40 (1966); A. M. Liquori, Q. Rev. Biophys., 2, 165 (1969).
29. Y. Y. Gotlied, Soviet Phys. Tech. Phys., 4, 465 (1959); ibid. 2, 637, (1957).
30. T. M. Birchstein and O. B. Ptitsyn, Conformation of Macro-molecules (Interscience Publishers, N.Y. 1966).
31. K. Nagai, J. Chem. Phys., 38, 924 (1963); 40, 2818 (1964); 45, 838 (1966).
32. J. E. Mark, J. Am. Chem. Soc., 88, 4354 (1966); 88, 3708 (1966); 89, 6829 (1967).
33. P. J. Flory, Statistical mechanics of chain molecules (Interscience Publishers, New York 1969).
34. R. M. Pitzer and W. W. Lipscomb, J. Chem. Phys., 39, 1995 (1963).
35. E. Clementi and R. D. Davies, J. Chem. Phys., 45, 2593 (1966)
36. W. Fink, D. C. Pan and L. C. Allen, J. Chem. Phys., 47, 895 (1967).
37. L. Pederson and K. Morokuma, J. Chem. Phys., 46, 3941 (1967).
38. A. Veillard, Chem. Phys. Letters, 3, 128 (1969); 3, 565 (1969).
39. E. Clementi, Physics of Electronic and Atomic Collisions, pp. 399-426 North-Holland Publishers (1971).
40. G. C. Lie and E. Clementi, J. Chem. Phys., 60, 3005 (1974).
41. a. O. Matsuoka, C. Tosi and E. Clementi, Biopolymers, 17, 33 (1978).
    b. C. Tosi, O. Matsuoka and E. Clementi, Biopolymers, 17, 51 (1978).
    c. C. Tosi, O. Matsuoka and E. Clementi, Biopolymers, 17, 67 (1978).
    d. C. Tosi, E. Pescatori and E. Clementi, Biopolymers, 18, 203 (1979).
42. S. R. Niketic and K. Rasmusse, Lecture Notes in Chemistry, Vol. 3, Springer-Verlag, Berlin (1977). The Constitent Force Field.
43. J. P. Malrieu, Electronic Structure Calculation (G. A. Segal, Ed.) Plenum Press, New York, 1977.
44. J. Hermans, D. R. Ferro, J. E. McQueen, S. C. Wet, in "Environmental Effects On Molecular Structure And Properties", B. Pulmann, Ed., Reidel, Dordrecht, 1976, p. 459.
45. G. Corongiu and E. Clementi (unpublished data).
46. G. Bolis, M. Ragazzi, D. Salvaderi, D. R. Ferro and E. Clementi, Int. J. Quant. Chem., 14, 815 (1978).
47. J. Drenth, J. N. Jansonious, R. Koexoek, H. M. Swen, B. G. Wolthers, Nature (London), 218, 929 (1968).
48. J. Drenth, H. M. Swen, W. Hoogenstraaten, L. A. A. Sluyterman, Proc. K. Akad. Wet. Amsterdam, B78, 1C4 (1975).
49. J. Drenth, K. H. Kalk, H. M. Swen, Biochemistry, 15, 3731 (1976).
50. S. Scheiner, D. A. Kleier, W. N. Lipscomb, Proc. Nat. Acad. Sci. U.S.A., 72, 2660 (1975).
51. R. Broer, P.Th. Van Duijnen, W. C. Nieuwpoort, Chem. Phys. Lett., 42, 525 (1976).
52. F. A. Momany, R. F. McGuire, A. W. Burgess, H. A. Scheraga, J. Phys. Chem., 79, 2361 (1975).
53. E. Clementi, J. Chem. Phys. 46, 3851 (1967); ibid. 47, 2323 (1967).
54. E. Clementi, J. Chem. Phys., 46, 3842 (1967).
55. R. S. Mulliken, Chem. Revs., 9, 347 (1931); Rev. Mod. Phys., 4, 1 (1932); see also I. Wigner, E. E. Witmer, Z. Physik, 51, 859 (1928) and F. Hund, Z. Physik, 42, 93 (1927).
56. G. Herzberg, Z. Physik, 57, 616 (1929).
57. R. F. W. Bader, T. T. Nguyen-Dang, Y. Tal, J. Chem. Phys., 70, 4316 (1979).
58. K. Collard, G. G. Hall, Int. J. Quantum Chem. 12, 623 (1977).
59. W. Heitler, G. Rumer, Z. Phys., 68, 12 (1931). See, in addition, J. H. Van Vleck, J. Chem. Phys., 1, 219 (1933).
60. R. S. Mulliken, J. Chem. Phys., 2, 782 (1934).

61. H. H. Voge, J. Chem. Phys., $\underline{4}$, 581 (1936).
62. C. A. Coulson, Valence (Oxford Press - Oxford) 1961.
63. E. Clementi, A. Routh, Int. J. Quantum Chem., $\underline{6}$, 525 (1972).
64. E. Clementi (to be published).
65. E. Clementi, "Revision de la Computation Mechanico-Quantica de Atoms y Moleculas", p. 246, Special IBM Technical Report, July 1974.
66. This basis set has been published by F. Van Duijneveldt, Department of Chemistry, University of Utrecht (see IBM Technical Report, R. J. 045 - December 10, 1971). See also L. Gianolio, R. Pavani, E. Clementi, Gazz. Chim. Ital., $\underline{108}$, 181 (1978).
67. W. C. Ermler, C. W. Kern, J. Chem. Phys., $\underline{58}$, 3458 (1973). We refer to this work for an extended discussion on the Hartree-Fock limit for the total energy, for the experimental geometry and dissociation energy values, and for an estimate of the extra-molecular correlation energy.
68. H. Popkie, E. Clementi, J. Chem. Phys., $\underline{57}$, 1077 (1972).
69. E. Clementi, D. Raimondi, J. Chem. Phys., $\underline{38}$, 2686 (1963); E. Clementi, "Tables of Atomic Functions:, IBM J. Res. Dev. Suppl., $\underline{9}$, 2 (1965); E. Clementi, C. Roetti, "Atomic Data and Nuclear Data Tables" Vol. $\underline{14}$, No. 3 and No. 4, Academic Press, New York (1974).
70. E. Clementi, Lecture Notes in Chemistry, Vol. II, Springer-Verlag, New York, Heidelberg, 1976.
71. E. Clementi, Bull. Soc. Chim. Belg., $\underline{85}$, 969 (1976).
72. E. Clementi, J. Chem. Phys. $\underline{46}$, 4731 (1967).
73. G. Giunchi, E. Clementi, M. E. Ruiz-Vizcaya, O. Novaro, Chem. Phys. Lett., $\underline{49}$, 8 (1977).
74. O. Novaro, E. Blaisten-Barojas, E. Clementi, G. Giunchi, M. E. Ruiz-Vizcaya, J. Chem. Phys., $\underline{68}$, 2337 (1978).
75. G. Corongiu and E. Clementi, Gazz. Chim. It. $\underline{108}$, 273 (1978).
76. E. Clementi, G. Corongiu, B. Jonsson and S. Romano, FEBS, $\underline{100}$, 313 (1979); J. Chem. Phys., $\underline{72}$, 260 (1980).
77. E. Clementi, G. Corongiu, B. Jonsson and S. Romano, Montedison Tech. Rep. DDC-802 (December 1978) and Gazz. Chim. It. (in press).
78. Y. Pocker, S. Sarkanen, Advances in Enzymology, $\underline{47}$, 149 (1978).
79. E. Clementi, R. Scordamaglia, and F. Cavallone, J. Am. Chem. Soc., $\underline{99}$, 5531 (1977) and L. Carozzo, G. Corongiu, C. Petrongolo, E. Clementi, J. Chem. Phys., $\underline{68}$, 787 (1978).
80. R. Scordamaglia, F. Cavallone and E. Clementi, J. Am. Chem. Soc., $\underline{99}$, 5545 (1977).
81. G. Bolis and E. Clementi, J. Am. Chem. Soc., $\underline{99}$, 5550 (1977).
82. M. Ragazzi, D. R. Ferro and E. Clementi, J. Chem. Phys., $\underline{70}$, 1040 (1979)
83. G. Corongiu, E. Clementi, E. Pretsch and W. Simon, J. Chem. Phys., $\underline{70}$, 1266 (1979).
84. G. Corongiu, E. Clementi, E. Pretsch and W. Simon, J. Chem. Phys., (in press); G. Ranghino, E. Clementi (to be published).
85. H. Popkie, H. Kistenmacher and E. Clementi, J. Chem. Phys., 58, 1689 (1973).
86. H. Kistenmacher, H. Popkie and E. Clementi, J. Chem. Phys., $\underline{58}$, 5627 (1973).
87. H. Kistenmacher, H. Popkie and E. Clementi, J. Chem. Phys., $\underline{59}$, 5842 (1973).
88. H. Kistenmacher, H. Popkie and E. Clementi, J. Chem. Phys., $\underline{60}$, 4455 (1974).
89. a. H. Popkie, H. Kistenmacher and E. Clementi, J. Chem. Phys., $\underline{59}$, 1325 (1973).
    b. G. C. Lie and E. Clementi, J. Chem. Phys., $\underline{62}$, 2195 (1975).
90. G. C. Lie, M. Yoshimine and E. Clementi, J. Chem. Phys., $\underline{64}$, 2314 (1976) and O. Matsuoka, M. Yoshimine, E. Clementi, J. Chem. Phys., $\underline{64}$, 1351 (1976).

91. E. Clementi and G. Corongiu. Pair potentials for amino acids-amino acids (to be published).
92. E. Clementi, D. R. Davies, J. Comput. Phys., $\underline{1}$, 223 (1968).
93. L. Gianolio, R. Pavani and E. Clementi, Gazz. Chim. It., $\underline{108}$, 181 (1978).
94. L. Gianolio and E. Clementi, Gazz.Chim. It. (in press).
95. S. F. Boys and F. Bernardi, Molecular Physics $\underline{19}$, 553 (1970).
96. W. Kolos, Theor. Chim. Act., $\underline{51}$, 219 (1979).
97. W. A. Latham, G. R. Pack, K. Morokuma: J. Am. Chem. Soc., $\underline{97}$, 6624 (1975).
98. F. London: Z. Physik. Chemie (B) $\underline{11}$, 222 (1930), Trans. Faraday Soc., $\underline{33}$, 8 (1937).
99. P. Claverie: "Elaboration of Approximate Formulas for the Interactions between Large Molecules": in "Intermolecular Interactions: from Diatomics to Biopolymer", Ed., B. Pullman, J. Wiley, New York 1978.
100. Landolt-Bornstein, Zahlenwerte und Funktionen; J. Bartel, et al, Eds., Springer Verlag, Berlin 1951, Vol. I/3, p. 513.
101. C. G. Le Fevre, R. J. W. Le Fevre: Rev. Pure Appl. Chem. $\underline{5}$, 261 (1955).
102. G. D. Zeiss, W. J. Meath: Mol. Phys. $\underline{30}$, 161 (1975).
103. A. Dalgarno: Adv. Chem. Phys. $\underline{12}$, 143 (1967).
104. J. T. Egan, T. J. Swissler, R. Rein: Int. J. Quantum Chem., Quantum Biology Symp. $\underline{1}$, 71 (1974).
105. Reference 100, p. 511.
106. W. Kolos, G. Ranghino, O. Novaro, and E. Clementi, Int. J. Quantum Chem. $\underline{17}$, 429 (1980).
107. W. Kolos, G. Corongiu, E. Clementi, Int. J. Quantum Chem. (in press).
108. A. D. Buckingham, Adv. Chem. Phys. $\underline{12}$, 107 (1967).
109. W. B. Neilsen and R. G. Gordon, J. Chem. Phys. $\underline{58}$, 4149 (1973).
110. R. M. Herman, J. Chem. Phys. $\underline{44}$, 1346 (1966).
111. K. T. Tang, J. M. Norbeck and P. R. Certain, J. Chem. Phys. $\underline{64}$, 3063 (1976).
112. N. J. Bridge and A. D. Buckingham, Proc. Roy. Soc. (London) A$\underline{295}$, 334 (1966).
113. G. Duguette, T. M. Ellis, G. Scoles and R. O. Watts, J. Chem. Phys. $\underline{68}$, 2544 (1978).
114. Quoted in Reference 113.
115. R. T. Pack, J. Chem. Phys. $\underline{64}$, 1659 (1976).
116. E. A. Mason and L. Monchick, J. Chem. Phys. $\underline{35}$, 1676 (1961).
117. S. L. Holmgren, M. Waldman and W. Klemperer, J. Chem. Phys. $\underline{69}$, 1661 (1978).
118. S. E. Novick, P. Davies, S. J. Harris and W. Klemperer, J. Chem. Phys. $\underline{59}$, 2273 (1973).
119. S. E. Novick, K. C. Janda, S. L. Holmgren, M. Waldman and W. Klemperer, J. Chem. Phys. $\underline{65}$, 1114 (1976).
120. J. M. Farrar and Y. T. Lee, Chem. Phys. Lett. $\underline{26}$, 428 (1974).
121. C. Tosi, R. Scordamaglia, E. Clementi, D. H. Wertz and H. A. Scheraga, to be published.
122. F. B. van Duijneveldt, IBM Technical Report RJ945, December 10, 1971.
123. W. Kolos, private communications.
124. E. Clementi and H. Popkie, J. Am.Chem. Soc. $\underline{94}$, 4957 (1972).
125. W. J. Hehre, R. F. Stewart and J. A. Pople, J. Chem. Phys. $\underline{51}$, 2657 (1969).
126. C. R. A. Catlow, A. H. Harker and M. R. Hayns, J. Chem. Soc., Faraday Trans. II $\underline{71}$, 275 (1975).
127. W. Kolos, Theo. Chim. Acta (Berl.) (in press).
128. A. Dalgarno, Adv. Chem. Phys. $\underline{12}$, 143 (1967).
129. P. Matthews and B. Smith, Mol. Phys. $\underline{32}$, 1719 (1976).

130. A. Buckingham, Disc. Faraday Soc. 40, 232 (1965); H. N. W. Lekkerkerker, P. Coulon and R. Luyckx, J. Chem. Soc., Faraday Trans. 2, 73, 1328 (1977).
131. H. Kistenmacher, H. Popkie and E. Clementi, J. Chem. Phys. 61, 799 (1974).
132. J. W. Kress, E. Clementi, J. J. Kazak and M. E. Schwartz, J. Chem. Phys. 63, 3907 (1975).
133. D. Hankins, J. W. Moskowitz and F. H. Stillinger, J. Chem. Phys. 53, 4544 (1970).
134. J. E. Del Bene and J. A. Pople, J. Chem. Phys. 58, 3605 (1973).
135. J. E. Del Bene, J. Chem. Phys. 55, 4633 (1971).
136. H. Kistenmacher, G. C. Lie, H. Popkie and E. Clementi, J. Chem. Phys. 61, 545 (1974).
137. E. Clementi, W. Kolos, G. C. Lie, and G. Ranghino, Int. J. Quantum Chem. 17, 377 (1980).
138. W. Kolos, F. Nieves and O. Novaro, Chem. Phys. Lett. 41, 431 (1976); O. Novaro and W. Kolos, J. Chem. Phys. 67, 5066 (1977).
139. O. Novaro and V. Beltran-Lopez, J. Chem. Phys. 56, 815 (1972); O. Novaro and F. Yanez, Chem. Phys. Lett. 30, 60 (1975); W. Kolos and A. Les, Int. J. Quantum Chem. 6, 1101 (1972).
140. E. Clementi in "Physics of Electronic and Atomic Collisons", VII IOPEAC, 1971 (North Holland).
141. E. Clementi, H. Kistenmaker, W. Kolos and S. Romano, Theor. Chim. Acta (in press).
142. B. M. Axilrod and E. Teller, J. Chem. Phys. 11, 255 (1943).
143. J. E. Egan, J. T. Swissler and R. Rein, Int. J. Quant. Chem. Quantum Biol. Symp. 1, 71 (1974).
144. J. A. Barker and R. O. Watts, Chem. Phys. Letters, 3, 144 (1969).
145. O. Matsuoka, M. Yoshimine, E. Clementi, J. Chem. Phys., 64, 1351 (1976).
146. M. Mezei, S. Swaminathan, D. L. Beveridge, J. Chem. Phys., 71, 3366 (1979).
147. M. Mezei, S. Swaminathan, D. L. Beveridge, J. Am. Chem. Soc., 100, 3255 (1978).
148. S. Romano and K. Singer, Mol. Phys., 37, 1765 (1979).
149. P. A. Lebwohl and G. Lashar, Phys. Rev., A6, 426 (1972).
150. G. M. Torrie and G. P. Valleau, Chem. Phys. Letters, 28, 578 (1974).
151. J. P. Valleau and D. N. Card, J. Chem. Phys., 57, 5457 (1972).
152. G. M. Torrie and J. P. Valleau, J. Chem. Phys. 66, 1402 (1977).
153. B. Widom, J. Chem. Phys. 39, 2808 (1963).
154. Z. Slanina, Int. J. Quantum Chem., 16, 79 (1979).
155. G. Corongiu and E. Clementi, J. Chem. Phys. 69, 4885 (1978).
156. R. Barsotti and E. Clementi, Theor. Chim. Acta, 42, 101 (1977).
157. E. Clementi and R. Barsotti, Chem. Phys. Letters, 59, 21 (1978).
158. J. D. Bernal and R. H. Fowler, J. Chem. Phys. 1, 515 (1933).
159. E. J. Verwey, Rec. Trav. Chim. 61, 127 (1942).
160. W. E. Morf and W. Simon, Helv. Chim. Acta 54, 794 (1971).
161. S. Romano and E. Clementi, Gazz. Chim. It., 108, 319 (1978).
162. W. S. Benedict, N. Gailar, F. K. Plyler, J. Chem. Phys., 24, 1139 (1956).
163. S. Romano and E. Clementi, Int. J. Quant. Chem. (in press).
164. P. G. Jonsson, A. Kvick, Acta Crystallogr., Sect. B. 28, 1827 (1972).
165. P. C. Fantucci and L. Gianolio, Istituto Richerche "G. Donegani", Montedison SpA; Technical Report DDC - 803, December 1978.
166. G. Ranghino and E. Clementi, Gazz. Chim. It., 108, 170 (1978), see also G. Ranghino, R. Scordamaglia and E. Clementi, Chem. Phys., Lett., 49, 218 (1977).
167. J. Moult, A. Yonath, unpublished data.
168. A. T. Hagler, J. Moult, Nature, (London) New Biol., in press.
169. G. Bolis, Dissertation, University of Milano, Dept. of Physics (1977).

170. T. Imoto, L. N. Johnson, A. C. T. North, D. C. Philips, J. A. Rupley, in "The Enzymes", Vol. 7, 3rd ed., P. Boyer ed., Academic Press, New York, 1972, pp. 665-862.
171. J. Moult, A. Yonath, W. Traub, A. Smilansky, A. Podjarny, D. Rabinovich, A. Saya, J. Mol. Biol., $\underline{100}$, 179 (1976).
172. J. A. McCammon, B. R. Gelin, M. Karplus and P. G. Wolynes, Nature (London) New Biol., $\underline{262}$, 324-326 (1976); J. A. McCammon, B. R. Gelin, M. Karplus, ibid., $\underline{267}$, 585 (1977).
173. S. Romano and E. Clementi, Gazz. Chim. Ital. $\underline{108}$, 319 (1978); S. Romano and E. Clementi, Int. J. Quantum Chem., $\underline{14}$, 839 (1978).
174. This fact is experimentally well-known. The reported value of the activation energy is obtained from ab-initio computations by E. Jonsson, G. Karlstrom, H. Wennerstrom, S. Forsen, B. Roos, and J. Almlof, J. Am. Chem. Soc., $\underline{99}$, 4628 (1977) and references therein given.
174. E. Magid and B. O. Turbeck, Biochim. Biophys. Acta., $\underline{165}$, 512 (1968).
175. S. Lindskog, L. E. Henderson, K. K. Kannan, A. Liljas, P. O. Nyman and B. Standberg in: The Enzymes, Vol. 5, Page 587, Academic Press, London (1971).
176. K. K. Kannan, B. Notstrand, K. Fridborg, S. Lovgren, A. Ohlsson and M. Petef, Proc. Natl. Acad. Sci. USA, $\underline{72}$, 51 (1975). It is pleasure to thank Dr. K. K. Kannan for providing the X-ray coordinates.
177. B. Jonsson, G. Karlstrom and H. Wennerstrom, J. Am. Chem. Soc., $\underline{100}$, 1658 (1978).
178. D. Demoulin, A. Pullman and B. Sarkar, J. Am. Chem. Soc. $\underline{99}$, 8498 (1977), see also D. Demoulin, and A. Pullman, Theor. Chim. Acta., $\underline{49}$, 161 (1978).
179. E. Clementi, G. Corongiu, B. Jonsson and S. Romano, Montedison (Novara, Italy) Tech. Report DDC-802, December 1978 contains the coordinates of the 27 amino acids.
180. I. Bertini, G. Canti, C. Luchinat and Scozzafava, J. Am. Chem. Soc., $\underline{100}$, 4837 (1978).
181. H. Ohtaki, T. Yamaguchi and M. Maeda, Bull. Chem. Soc. Jpn., $\underline{49}$, 701 (1976).
182. J. O.M. Boakris and A. K. N. Reddy, Modern Electrochemistry, Vol. 1, Plenum Press, New York (1970).
183. X. Gmelin, "Handbuch det anorganischen Chemie" 8 Auflage, $\underline{32}$, 720 (1956).
184. E. Clementi, G. Corongiu and F. Lelj, J. Chem. Phys. $\underline{70}$, 3726 (1979).
185. E. Clementi and G. Corongiu, Gazz. Chim. It. $\underline{108}$, 687 (1978).
186. E. Clementi, J. M. Andre, M. Klint and D. Hahn, Acta Phys. Acad. Sci. Hung. $\underline{27}$, 493 (1969).
187. E. Clementi, J. Mehl and W. Von Niessen, J. Chem. Phys. $\underline{54}$, 508 (1971).
188. R. Rein, "Intermolecular Interactions" pp. 307-362, B. Pullman Ed., John Wiley & Sons, New York (1978).
189. E. Clementi and G. Corongiu, Biopolymers, $\underline{18}$, 2431 (1979).
190. E. Clementi and G. Corongiu, Gazz. Chim. It., $\underline{109}$, 201 (1979).
191. R. Fieldman, Atlas Macromolecules, document 13.1.1.1.1 (1976), Nat. Inst. Health, Bethesda, Maryland, U.S.A.
192. E. Clementi and G. Corongiu, Int. J. Quant. Chem., $\underline{16}$, 897 (1979); E. Clementi and G. Corongiu, Chem. Phys. Lett. $\underline{60}$, 175 (1979).
193. R. Fieldman, Atlas of Macromolecules, document 13.2.1.1.1 (1976), Natl. Inst. Health, Bethesda, Maryland, U.S.A.
194. E. Clementi and G. Corongiu, J. Chem. Phys. $\underline{72}$, 3979 (1980).
195. M. Falk, K. A. Hartman and R. C. Lord, J. Am.Chem. Soc., $\underline{84}$, 3843 (1962); ibid. $\underline{85}$, 387 (1963); ibid. $\underline{85}$, 391 (1963).
196. A. Rupprecht and B. Forslind, Biochim. Biophys. Acta., $\underline{204}$, 304 (1970).

197. J. E. Hearst and J. Vinograd, Proc. Natn. Acad. Sci., U.S.A., <u>47</u>, 825 (1961); ibid <u>47</u>, 999 (1961); ibid. <u>47</u>, 1005 (1961).
198. B. Wolf and S. Hanlon, Biochemistry <u>14</u>, 1661 (1975).
199. M. J. B. Tunis and J. E. Hearst, Biopolymers <u>6</u>, 1325 (1968); ibid. <u>6</u>, 1345 (1968).
200. I. D. Kuntz, T. S. Brassfield, G. D. Law and G. V. Purcell, Science, N.Y., <u>163</u>, 1329 (1969).
201. P. L. Privalov, O. B. Ptitsyn and T. M. Birshtein, Biopolymers, <u>8</u>, 559 (1969).
202. J. Texter, Prog. Biophys. Molec. Biol., <u>33</u>, 83 (1978).
203. G. Corongiu and E. Clementi (Na^+ interaction potential with sugar, phosphates and bases; unpublished data).
204. E. Clementi and G. Corongiu (the data of sections 5.12 and 5.18 have been presented at the Am. Chem. Soc. Meeting, Houston, Texas, U.S.A., March 24, 1980 and at the International Symposium on Quantum Biology, Palm Beach, Florida (March 4, 1980).
205. S. Romano, G. Corongiu and E. Clementi (unpublished data).

A. F. Williams

A Theoretical Approach to Inorganic Chemistry

1979. 144 figures, 17 tables. XII, 316 pages
ISBN 3-540-09073-8

This book is intended to outline the application of simple quantum mechanics to the study of inorganic chemistry, and to show its potential for systematizing and understanding the structure, physical properties, and reactivities of inorganic compounds. The considerable development of inorganic chemistry in recent years necessitates the establishment of a theoretical framework if the student is to acquire s sound knowledge of the subject. An effort has been made to cover a wide range of subjects, and to encourage the reader to think of further extensions of the theories discussed. The importance of the critical application of theory is emphasized, and, altrough the book is concerned chiefly with molecular orbital theory, other approaches are discussed. The book is intended for students in the latter half of their undergraduate studies.

Contents: Quantum Mechanics and Atomic Theory. – Simple Molecular Orbital Theory. – Structural Applications of Molecular Orbital Theory. – Electronic Spectra and Magnetic Properties of Inorganic Compounds. – Alternative Methods and Concepts. – Mechanism and Reactivity. – Descriptive Chemistry. – Physical and Spectroscopic Methods. – Appendices. – Subject Index.

Springer-Verlag
Berlin
Heidelberg
New York

Hermann Hartmann

**Die chemische Bindung
Drei Vorlesungen für Chemiker**

3. Auflage 1971. 61 Abbildungen. V, 109 Seiten
ISBN 3-540-03145-6

„Den meisten Physikern fehlen die eingehenden chemischen Grundlagen, den Chemikern die zum Verständnis der chemischen Bindung nötigen mathematischen und theoretisch-physikalischen Voraussetzungen. Hartmann hat mit Erfolg einen Versuch gemacht, diese Schwierigkeiten zu überbrücken. Das Büchlein enthält den Stoff von drei Vorlesungen. Der Autor geht von den wichtigsten atomistischen Vorstellungen aus, behandelt die Energiegehalte der Atome, die Quantelung der Energiezustände, die Welleneigenschaften der Korpuskeln, die Grundzüge der Quantenmechanik, die Elektronenschalen und ihre Besetzung, das Energieschema des Wasserstoff-Ions, verschiedene Bindungsarten, Convalenz und Elektrovalenz, die Bindung im Wasserstoffmolekül, Elektronenzustand und Valenzwinkel beim Kohlenstoffatom, die einfachsten Kohlenstoffverbindungen, die Unterschiede zwischen Atombindung, Ionenbindung und metallischer Bindung und die Ionenradien in verschiedenen Gittern.

Hartmann setzt wenig mathematische und theoretisch-physikalische Kenntnisse voraus. Trotzdem wird erstaunlich viel in der knappen Darstellung von 100 Seiten behandelt. Wenn er sich darin in erster Linie an die Chemiker wendet, ist es doch auch für den Physiker hochinteressant und lehrreich und kann daher beiden bestens empfohlen werden."

aus: Physikalische Blätter

Springer-Verlag
Berlin
Heidelberg
New York